# Genetics of
# Sex Differentiation

# Genetics of
# Sex Differentiation

URSULA MITTWOCH

*The Galton Laboratory*
*Department of Human Genetics and Biometry*
*University College London*
*London, England*

ACADEMIC PRESS New York and London 1973
*A Subsidiary of Harcourt Brace Javonovich, Publishers*

ACADEMIC PRESS, INC.
111 Fifth Avenue, New York, New York 10003


*United Kingdom Edition published by*
ACADEMIC PRESS, INC. (LONDON) LTD.
24/28 Oval Road, London NW1


**Library of Congress Cataloging in Publication Data**

Mittwoch, Ursula.
 Genetics of sex differentiation.

 Bibliography: p.
  1. Sex chromosomes. 2. Sex–Cause and determination. 3. Sex chromosome abnormalities. I. Title.
[DNLM: 1. Cytogenetics. 2. Sex chromosomes.
3. Sex determination. WQ 206 M685g 1973]
QH431.M537  575.1  72–9334
ISBN 0–12–501040–0

# Contents

*v*

## Chapter 4   **Heterochromatin**

## Chapter 5   **The Nature of Sex Differentiation with Special Reference to Vertebrates**

## Chapter 6   **Sex Determination in Man and Other Mammals**

## Chapter 7 Genes, Chromosomes, Growth, and Sex

# Preface

The role played by chromosomes in the development of male and female characteristics is a topic of exceptional interest. It has been almost seventy years since sex chromosomes were discovered, providing the first real answer to the age-old question as to the cause of sex differentiation. Nevertheless, the hopes of early geneticists to explain the determination of sex in terms of classical Mendelian genetics were never fulfilled. In spite of the accumulation of much pertinent data, no solid evidence could be adduced for the existence of either male- or female-determining genes in *Drosophila* or other organisms. As a result, efforts to unravel the genetics of sex determination were abandoned in favor of investigations which seemed to fit more readily into the Mendelian framework.

This is an opportune time to reinvestigate the problem of sex differentiation for several reasons. The unprecedented progress in the cytogenetics of man and other mammals has focused attention on the relationship between chromosomes and normal, as well as abnormal, sexual development. The large number of sex chromosome abnormalities that have come to light in our own species have forced us to take a closer look at the manifold developmental processes that occur from sex determination (at conception) to sexual maturity. But progress in clinical genetics has been matched by equally far-reaching discoveries of a more fundamental nature. It has become apparent that the chromosomes of higher organisms contain a large amount of DNA and that some of it does not function in accordance with the genetic code. Furthermore, dramatic advances in our knowledge of this nongenic DNA are continually being made with the aid of newly developed techniques. It is clear that there is a striking correspondence between this part of the DNA and the chromosomal regions that have long been regarded as heterochromatic. It has also long been known that sex chromosomes are particularly rich in heterochromatin.

The purpose of this book is to bring together evidence that the sex chromosomes may affect the rates at which cells divide and that the process of sex differentiation is based on differences in growth rates during development. This has necessitated discussion on a rather wide variety of topics, and no attempt has been made to cover any one exhaustively. Although some of the ideas mentioned are themselves in an early stage of development and no doubt in need of modification, it is hoped that the approach I have chosen—looking beyond the formal gene–phenotype relationship and emphasizing the dynamic relationship between chromosomes and growth—will lead to a better understanding of the role of chromosomes in the development of sexual and other characteristics, i.e., those which are basically of a quantitative nature. The realization that chromosomes, in addition to carrying genes determining chemical specificity, contain other regions which control the rates of cell division and growth should help to ally cytogenetics with embryology and evolution and generally shed light on the interaction between nature and nurture. The process of sex differentiation provides an excellent system to test such correlations.

It is a pleasure to acknowledge the help of a number of colleagues who kindly read individual chapters in the manuscript and made valuable comments: A. Anders, F. Anders, M. J. Fahmy, O. G. Fahmy, C. E. Ford, G. R. Fraser, W. Landauer, R. C. Lewontin, and C. A. B. Smith. I am also grateful to those authors and publishers who have permitted me to use previously published illustrations and particularly to those who have contributed original photographs to be included in this book; they have been acknowledged in the text.

I should like to thank Miss Ruth Lang and Mrs. Lilian Nutter for their patience and care in typing the manuscript, the staff of

Collings Design Group for their accuracy and speed in preparing the diagrams, and the staff of Academic Press for their help in seeing the book through production.

Some of the experimental work described in this book was supported by a grant from the Science Research Council.

URSULA MITTWOCH

# Genetics of
# Sex Differentiation

# Chapter 1

## Classical Genetics of Sex Differentiation

### I. Introduction

The problem of what determines sex has been debated since ancient times and has given rise to literally hundreds of hypotheses. It was not, however, until the beginning of the twentieth century that an answer, which had any relation to the underlying facts, was found. In 1899, the French biologist Cuénot, who a few years later was to be the first to demonstrate Mendelian inheritance in a mammal, wrote the following: "It is rather humiliating to state that as regards man and other mammals no advance has been made since the time of the predecessors of Aristotle, even though a considerable amount of work has been expended in trying to solve this problem [of sex determination]; evidently, the wrong approaches were chosen" (Author's translation).

Theoretically, sex could be determined at any one of three stages: (1) prior to fertilization by the constitution of the egg, the sperm having no effect on the sex of the zygote, (2) at the moment of fertilization, and (3) after fertilization through external influences on the developing embryo. During the second half of the nineteenth century the last view prevailed, and reports purporting to show that, in lower animals and in plants, the sex of the offspring could be varied by differences in nutrition were numerous. In man, the state of nutrition of the parents was thought to be important—there were those who thought that the better nourished parent transmitted his or her sex, while others held the opinion that the sex of the embryo was the opposite to that of the better nourished parent.

In the 1860's, much excitement was caused by a theory that the sex of the offspring was determined by the age of the egg at the time of fertilization. It was claimed that the sex of cattle could be selected depending on the point in the reproductive cycle at which a cow was allowed to be fertilized. Early fertilization was thought to result in female calves, while delayed fertilization of the egg was supposed to result in bull calves. Some investigators ascribed a corresponding effect to the age of the sperm, young sperm resulting in male, and older sperm in female, offspring. As it could be argued that by this means the sex which in the parental generation was numerically in the minority would increase its numbers, this system was thought to have an evolutionary advantage. The popular idea that there is an increase in the proportion of boys born during or after wars was explained based on this theory, although other possible causes, such as an improvement or worsening of nutrition, were also advanced. Last, such factors as the age of the father or the mother, the difference in age between the parents, the degree of inbreeding, or the season in which conception took place were thought to influence the sex at birth.

The question of what determines sex was unanswerable before the advent of Mendelian genetics and the development of the chromosome theory of heredity, that is to say, before the beginning of the twentieth century. Nevertheless, by the end of the nineteenth century some impressive data had been collected regarding the proportion of male and female births in man and domestic animals (Düsing, 1884; Janke, 1888). These data, which comprised many thousands of individuals, showed, in the first place, that in each species the ratio of male to female births is a characteristic figure, which can be varied only slightly, if at all, by environmental conditions (Henneberg, 1897). Therefore, such conditions could, at best, favor the production of one or the other sex but could not determine it. Second, it became clear that large samples are an essential prerequisite for drawing valid conclusions regarding sex ratios and that many of the earlier theories could be discounted for this reason. At the close of the century, Cuénot (1899) was able to dismiss unequivocally the theory that sex is determined solely by conditions operating during gestation. The stage was thus set for the discovery that sex is determined by the hereditary material.

However, the inheritance of sex differs in some respects from that of most other contrasting characters, and to understand the biological basis of sex determination it is necessary to appreciate certain fundamental developments which lead to an understanding of the role played by cells. It became evident first that the bodies of higher organisms, plants as well as animals, are composed of a very large number of individual cells, and second that new cells can originate only by the division of preexisting parent cells. These two principles made it possible to understand the facts of fertilization, which consists essentially of the fusion of two parental germ cells. When the spermatozoon fuses with the egg cell, a zygote cell is formed which, by repeated divisions, forms all the cells required by the new organism.

## II. The Structure of Cells

It became apparent that there is a striking similarity of the basic structure in all higher animals and plants. Each cell consists of a nucleus surrounded by cytoplasm. When a cell divides, the nucleus undergoes a series of characteristic processes during which the chromosomes become visible as individual entities. The number of chromosomes is characteristic for any given species and, with very few exceptions, is constant in different cells of the body. During cell division the chromosomes are longitudinally symmetrical and finally divide longitudinally, so that each daughter cell receives one-half of each chromosome.

Egg and sperm cells differ very considerably in size, the egg being a much larger cell, carrying a large amount of cytoplasm (Chapter 6). By contrast, the chromosomal endowments of egg and sperm cells are equivalent. This fact, as well as the exact distribution of chromosomes during cell division, suggested that it is the chromosomes which are the bearers of those hereditary characteristics which are passed on equally through the father and the mother.

The diameter of a typical cell is of the order of 20 $\mu$m, or one-fiftieth of a millimeter. To study the chromosomes in any detail requires the use of a light microscope at the maximum possible development of optical perfection. Such instruments became available in the last quarter of the nineteenth century. Accordingly, all the basic facts about chromosomes which have just been summarized were known at the close of the century. It was, however, not until the 1950's that it was discovered that the actual duplication of the chromosomes occurs some time prior to cell division, when the chromosomes are not visible as individual entities (Howard and Pelc, 1953). This means that the chromosomes, which have been studied so assiduously by generations of cytologists are, in fact, structures which have duplicated in readiness for cell division (hence, the longitudinal symmetry). We still know very little about the structure and activity of ordinary chromosomes in non-dividing cells. It is as if the history of mankind had been written by a midwife, whose only knowledge of people was of women in labor.

### III. The Chromosomal Basis of Sex Determination

Regarding the chromosomal basis of sex determination, the nineteenth century seems to provide only few forerunners to present-day thought. In a letter written to Naegeli in 1870, Mendel mentioned his supposition that the sex of a dioecious plant may be determined in its primordial cell. This supposition was based on an experiment crossing a dioecious species (i.e., male and female flowers borne on different plants) of *Silene* with that of a hermaphrodite; this was said to result in 152 female and 51 male plants, i.e., a 3:1 ratio. The data were quite fortuitous, but the conclusion was basically correct. As it happened, Naegeli was no more interested in Mendel's idea of the inheritance of sex than in Mendel's ideas on heredity as a whole.

The earliest observable evidence of a sex chromosome actually appears to date from the year 1891, when Henking studied sperm development in the insect, *Pyrrhocoris apterus*. He observed that during the maturation divisions of the spermatocyte, a peculiar chromatin element, which was sometimes more darkly stained than the other chromosomes, lagged behind and during the second maturation division passed undivided to one of the daughter cells. Thus, two types of spermatozoa were formed, those which contained the chromatin element and those which did not. Henking thought of this structure as partly chromosome and partly nucleolus, and this is hardly surprising when one considers how much confusion the heterochromatic chromosomes wrought among subsequent generations of investigators. In his drawings, he labeled the perplexing element "X" and this letter has since become the foremost in the chromosome alphabet. In another insect, the "squash bug" *Anasa tristis*, Paulmier (1899) gave a detailed description of a special chromosome which, during the second meiotic division of the spermatocyte, passed undivided into one of the daughter nuclei.

Other investigators of insect spermatogenesis also commented on the peculiar chromatin element, but its significance remained obscure. Even when its true chromosomal nature was realized, there was a tendency to assume that it was a chromosome in the process of disappearing, since it failed to divide during the second meiotic division and is not present in all spermatozoa. A revolutionary idea was then advanced by C. E. McClung, of Kansas University. In a long article entitled "The Accessory Chromosome—Sex Determinant?," McClung (1902) wrote as follows: "Assuming it to be true that the chromatin is the important part of the cell in the matter of heredity, then it follows that we have two kinds of spermatozoa that differ from each in a vital matter. We expect, therefore, to find in the offspring two sorts of individuals in approximately equal numbers, under normal conditions, that exhibit marked differences in structure. A careful consideration will suggest that nothing but sexual characters thus divides the members of a species into two well-defined groups, and we are logically forced to the conclusion that the peculiar chromosome has some bearing upon this arrangement."

The idea that sex is determined by a chromosome thus originated as a daring leap into the unknown, for it by no means follows as a logical conclusion from the premises that the chromosomes are the bearers of hereditary characteristics and that spermatozoa differ in their chromosome contents. The fact that virtually nothing was known at the time about the chromosome constitution of males and females is reflected in McClung's secondary hypothesis that the accessory chromosome is male determining, apparently based on the assumption that if there is a special body, it should belong to the male. It is true that Sutton (1902), a student of McClung's, counted one chromosome *less* in oogonia and ovarian follicle cells of female grasshoppers (*Brachystola magna*) than in the spermatogonia of the male. However, the counts in the female get only a brief mention in the detailed and important paper on the chromosomes during spermatogenesis, and there is no evidence that these counts in the obviously unfavorable material of female gametogenesis served as the basis for theoretical considerations; the opposite was probably the case.

While McClung's article was awaiting publication in *Biological Bulletins,* he wrote a briefer paper summarizing his hypothesis, which appeared in the *Anatomischer Anzeiger* several months before the main paper (McClung, 1901). These two publications herald the birth of the chromosome theory of sex determination.

It is hardly surprising that McClung's hypothesis was not readily accepted by the scientific community. At a time when the chromosomal nature of the accessory chromosome was still in doubt, the assumption that it should play a key role in sex determination was indeed a sweeping generalization. Most investigators preferred less far-reaching hypotheses. Some thought that the sperm lacking the accessory chromosome was incapable of fertilization; others, as mentioned above, believed that the accessory chromosome itself was in the process of becoming extinct.

However, two investigators set out to examine the problem experimentally. Stevens (1905) studied spermatogenesis in five species of insects and found an accessory chromosome to be present in three of them. One species, the common meal worm, *Tenebrio molitor,* gave an important clue to the problem of sex determination, since the chromosomes of the female could be successfully examined. Both sexes had the same somatic chromosome number of 20, but in the males this included a small chromosome, while in the female all chromosomes were more or less the same size.

As a result of meiosis, two classes of sperm were formed, those with ten large chromosomes and those with nine large and one small one. Although meiosis in the female could not be observed, it was clear that the egg nuclei must be alike with regard to the number and size of chromosomes. Since there were twenty large chromosomes in female somatic cells and nineteen large and one small in male somatic cells, Stevens concluded that "it seems certain that an egg fertilized by a spermatozoon which contains the small chromosome must produce a male, while one fertilized by a spermatozoon containing ten chromosomes of equal size must produce a female." An essentially similar situation was found in most other species of Coleoptera, the order to which *Tenebrio* belongs, which were examined by Stevens (1906). However, in a few species an unpaired chromosome was present during spermatogenesis, and in these species the somatic chromosome number of the male was one less than in females (Fig. 1.1A and B).

It was at the same time that E. B. Wilson, at Columbia University, began his studies of the chromosomes of insects, and he too made the discovery that the dimorphism of spermatocytes may be brought about in two different ways: either a chromosome is present in one class and absent in the other, or both classes each possess a chromosome of different size (Wilson, 1905). However, it was Stevens who compared the chromosomes in males and females and first worked out the formal relationship of these chromosomes to the process of sex determination. In a footnote added to his paper after it went to press, Wilson summarized Stevens' results and added: "These very interesting discoveries, now in course of publication, . . . show that McClung's hypothesis may, in the end, prove to be well founded."

It must not be assumed that the experimental evidence in favor of a chromosomal basis of sex determination, striking though it was, was readily accepted. In insects, gynandromorphs are not uncommon and "even if one were to accept the presence of different chromosome numbers in the same individual, one would have to assume some very curious phenomena concerning cell divisions during the development of such organisms. There would have to be abnormal mitosis during the early cleavage division giving rise to daughter cells with different chromosome numbers. The situation

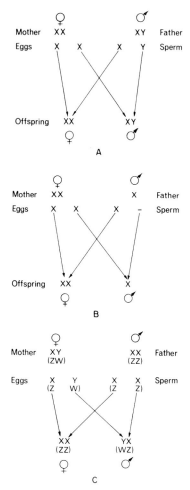

**Fig. 1.1.** Modes of sex determination. (A) Male heterogamety, Y chromosome present; (B) male heterogamety, Y chromosome absent; (C) female heterogamety.

becomes even more unlikely when one considers the so-called mixed gynandromorphs, where male and female characteristics occur side by side in all organs and tissues"* (Gross, 1906). In recent

* Author's translation.

years scientists were equally incredulous at the thought of chromosome mosaicism in mammals and, particularly, in man.

Wilson continued his investigations on the chromosomes of a large number of insect species, comparing the chromosomes of males and females and in mitosis and meiosis. As a result of his investigations, Wilson found in several species of the order Hemiptera (bugs) that the somatic chromosome number of the male is one less than that in the female and that there is a single unpaired chromosome during male meiosis, as had originally been observed in *Pyrrhocoris.* This chromosome, which had been given the names "chromatin nucleolus," "accessory chromosome," *chromosome spéciale,* "odd chromosome," "modified chromosome," "chromosome X," "heterochromosome," "allosome," "monosome," "idiochromosome," was called the "X chromosome" by Wilson (1909). The identity of the X chromosome in mitosis and meiosis could be proved in those species, such as *Protenor,* in which the X chromosome is the largest in the complement. As a result of meiosis, two types of sperm are formed, those with an X chromosome and those without an X chromosome. Since all egg cells contain an X chromosome, fertilization by an X-bearing sperm will result in a female, and fertilization by a sperm lacking an X chromosome will give rise to a male. Thus McClung's hypothesis was proved to be right in principle, although wrong in detail.

In other species of the Hemiptera, Wilson found a second type of sperm dimorphism identical with the one discovered by Stevens in the meal bug, *Protenor.* As before, the female has two X chromosomes and the male has one, but the male has an additional chromosome, which is not present in the female. Wilson called this the Y chromosome; usually it is smaller than the X chromosome. As a result of meiosis, the X and Y chromosomes separate, so that half the sperm contain an X and the other half a Y chromosome. As in the first scheme, sperm containing an X chromosome give rise to a female, but male-determining sperm contain a Y chromosome.

These findings were fully worked out in the first decade of the century and were summarized in an article by Wilson (1911) entitled "The Sex Chromosomes" in which he pointed out that he saw no reason why these chromosomes should not be given this name "even though we do not yet know precisely what is their causal relation to sex."

## IV. The Genetic Evidence

While cytologists were discovering the chromosomal basis of sex determination, the first decade of the twentieth century also saw the establishment of the foundations of Mendelian genetics. Whereas the cytology of sex chromosomes was almost an American monopoly, genetics, which developed out of the activities of plant and animal breeders, had strong roots in Europe. The principles of genetics were described by Mendel in a paper giving the results of hybridization experiments on the garden pea, *Pisum sativum,* and were published in 1866 in the *Proceedings of the Natural History of Brünn* (now Brno in Czechoslovakia, but at the time in the Austro-Hungarian Empire). This paper remained virtually unnoticed for 34 years, when it was rediscovered by three independent investigators, Correns in Germany, deVries in Holland, and Tschermak in Austria. The story of the rediscovery of this work, in the year 1900, has been described in recent years by Sturtevant (1965), Olby (1966), and Stern and Sherwood (1966).

Mendelian theory implies that the outward characteristics of living organisms, which differ in different individuals, are mediated by unit characters which Mendel called "factors" and which were subsequently called "genes" (Johannsen, 1909) and that these factors occur in pairs, of which only one member enters each germ cell. Therefore, the gametes of a hybrid organism are "pure" in as much as they carry only one of a pair of contrasting factors. To the "law of the purity of the gametes" Mendel added a second one, namely, the "law of independent assortment." The latter implies that if more than one pair of factors is involved, the individual members segregate in a random manner from each other prior to the formation of the germ cells.

Although, in his time, Mendel completely failed to interest his scientific colleagues in the idea of particulate units of heredity, by the year 1900 several investigators using similar techniques to Mendel's either arrived at the same idea or at least came very close to it. Hence the rediscovery of Mendel's paper. Its significance was now all the more apparent because of the discovery of the chromosomes and particularly of their segregation in meiosis prior to the formation of gametes.

Almost immediately after the rediscovery of Mendel's papers, Strasburger (1900) concluded that the sex of dioecious plants was not influenced by environmental conditions and was therefore likely to be determined by a hereditary quantity which he thought might be comparable to Mendel's factors. However, the difficulties inherent in attempts to assign genetic factors as causes of sex determination soon became apparent. Castle (1903) was the first to attempt to specify the mode of inheritance of the sexes. He assumed that both sexes are hybrids or heterozygotes, bearing the characters of both sexes, one in the dominant and the other in the recessive state. In the male, the female character is recessive, and vice versa. This scheme seemed to fit with the idea that each sex carried the characteristics of the opposite sex in a latent condition, shown by the occurrence of rudimentary organs characteristic of the other sex. There were, however, two disadvantages inherent in this scheme: The dominant and recessive conditions had to vary in the two sexes, and it required a system of differential fertilization, since gametes bearing one sex gene were assumed to be fertilized only by gametes bearing the other.

Altogether, the genetics of sex determination proved to be considerably less straightforward than that of many other hereditary traits. To begin with, in most animals the usual breeding techniques are not available, for the only type of mating possible is a cross between males and females, which in turn gives rise to males and females, and this does not advance the problem of the genetic basis of sex determination. However, Correns, one of the rediscoverers of Mendel's papers, evaded this dilemma by breeding plants instead of animals. In flowering plants, dioecism, i.e., the production of male and female organs by separate individuals, is the exception rather than the rule. Within the genus *Bryonia*, both dioecious and monoecious species occur. *Bryonia dioica* is dioecious, whereas *B. alba* is monoecious, i.e., male and female flowers are borne on the same plant. Correns (1907) made crosses between the two species with the following results.

B. *dioica* ♀ ✕ B. *alba* ♂ resulted in hybrid plants which were all female

B. *alba* ♀ ✕ B. *dioica* ♂ resulted in hybrid plants of which half were male and half female

Correns concluded that the sex of the plants must be determined at fertilization and that the male cells, or pollen grains, are of two types, male and female determining, while the eggs all have the same tendency toward sex determination. However, a few of the hybrids classified as females had a few rudimentary male flowers, and Correns was careful to point out that there was no question of the gametes transmitting the male or female tendencies in a pure condition. On the contrary, the gametes transmit the hermaphrodite condition, on which the tendency toward one sex is superimposed; this will develop while the other sex remains in a latent form. At that time it was clear, therefore, that the gametes are not pure for the factors of one or the other sex. Therefore, maleness and femaleness cannot be due to a pair of contrasting Mendelian factors.

In a latter work, Correns (1913) further emphasized the point that the hermaphrodite condition is not a hybrid between male and female characters, but that both sexes are basically hermaphrodite and that sex determination consists in developing one set of potentialities and suppressing the other set. Since each gamete transmits the tendency for both sexes, Correns argued that each individual must contain two male as well as two female genes and thus have the genetic constitution MFM'F'. Normally only one of these tendencies develop, while the other three remain latent. Because the potentialities for both sexes are present in both male- and female-determining cells, Correns (1913, 1928) postulated the existence of additional "sex determinants" (*Geschlechtsbestimmer*), which segregate in the germ cells and which normally determine the sex of an individual. These factors or genes were called *Realisatoren* by von Wettstein (1924) and this term has been widely used in the German literature (see Hartmann, 1956). We have thus arrived at two sets of sex-determining factors: (1) male and female potencies, or sex producers, which are common to all germ cells and (2) sex-determining genes, or sex differentiators, which segregate. With the advent of Mendelian genetics, many authors assumed that both sets represented pairs of genes, and notations such as MMFf and MMFF, for male and female, respectively, came into use (see Muller, 1932). Correns himself attempted to explain the process of sex determination in terms of a Mendelian backcross as follows: Females have the genetic consitution ff and produce germ cells, all of which are f, while males, of genetic constitution Mf, produce germ cells, half of which contain the M and half the f gene, so that on fertilization equal numbers of males and females are formed. This scheme implies that the gene for maleness must be dominant. The fundamental inconsistency inherent in this postulate, (which, in any case, could not apply to species which lack a Y chromosome) with the idea of the presence of male and female genes in both sexes has not been satisfactorily resolved until the present day.

## V. Female Heterogamety

We have seen that in many insects, as well as in the flowering plant *Bryonia dioica*, the males produce two types of games with regard to their sex chromosome constitution, whereas the sex chromosomes of female gametes are all the same. The males are therefore said to be "heterogametic" and the females "homogametic." The genetic basis of sex determination became further complicated when it was realized that in some organisms the situation is reversed.

The currant moth, *Abraxas grossulariata*, which is a pest of currant and gooseberry bushes, belongs to the order Lepidoptera. The species has a pale variety, the inheritance of which was worked out in the early days of genetics by Doncaster and Raynor (1906). Specimens of the pale variety taken in the wild are almost always female. When such a female is mated to a dark male, the offspring are all dark, indicating that this character is dominant. However, when a pale female is mated with one of her male offspring both pale and dark male and female offspring are produced. Finally, a dark female mated to a pale-colored male will have a dark male and pale female offspring. These results are explained by assuming that the gene for the color difference is carried on the X chromo-some, of which there are two in males and only one in females (Fig. 1.2). The characteristic pattern of inheritance of genes lo-cated on the X chromosome is known as "sex-linked."

The chromosomes of *Abraxas*, in common with those of other

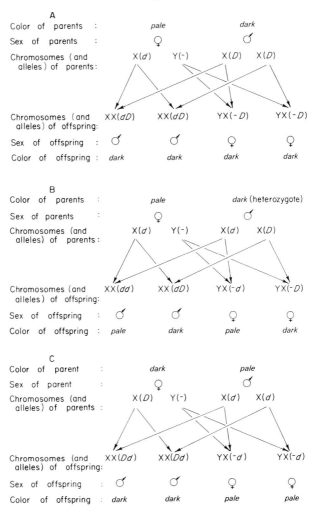

*Fig. 1.2.* Sex linkage in currant moth, *Abraxas* (female heterogamety).

Lepidoptera, have not proved favorable for cytological study; however, some cytological evidence for the existence of a chromosomal difference in different egg cells was presented by Doncaster (1914) and by Seiler (1914). Usually the chromosome number is the same in both sexes, indicating that a Y chromosome is probably present in females.

That birds are another class of animals in which the female is heterogametic also became known in the early years of genetics. It was pointed out by Spillman (1908) and subsequently confirmed by Morgan and Goodale (1912) that the inheritance of the gene for barred plumage in chickens is essentially similar to that for the pale variety in the currant moth. The plumage of the barred Plymouth rock fowl has a black and white striped appearance; an occasional black bird is usually a female. If a nonbarred female is mated to a barred male, all the offspring are barred: If the nonbarred female is mated to one of her male offspring, the offspring both males and females are nonbarred and barred. A barred female mated to a nonbarred male has barred male and nonbarred female offspring. The implication is clearly that the gene for "barred" is on the X chromosome (Hutt, 1949). The mode of inheritance of barred plumage in chickens is, therefore, just like that of dark color in *Abraxas*.

The chromosomes of birds were for a long time regarded as difficult cytological material. However, in recent years these difficulties have been overcome and the sex chromosomes have been unequivocally demonstrated in a number of species. It was shown by Ohno (1961) that in the fowl, *Gallus domesticus*, the female has one X chromosome and the male two; a Y chromosome was shown to be present in females by Fréderic (1961) and by Owen (1965).

In animals with female heterogamety, the unpaired sex chromosome is commonly referred to as W and the paired sex chromosomes of the male as ZZ. Obviously, this difference in terminology can do no more than indicate which sex has the unequal sex chromosomes. There is no question of homology between the X and Y chromosomes of men and insects or of the Z and W chromosomes of birds and Lepidoptera. These letters are merely symbols for paired and unpaired sex chromosomes. In 1932, Winge pointed out that the ZW was an unnecessary complication, and, it must be admitted that in the 40 years that have intervened, very little progress has been made as to the function of sex chromosomes in their role as sex determiners. The simpler XY terminology is still adequate for most purposes, and it is preferred in this text, regardless of whether the male or the female is the heterogametic sex. Apart from simplicity, a uniform terminology has the advantage of not prejudging the issue regarding the nature and function of the various sex chromosome constitutions.

The existence of male and female heterogamety provides a challenge to elucidate the role of the sex chromosomes in the development of the sex difference. Once this aim is achieved, a terminology will undoubtedly emerge which will reflect the true nature of the sex chromosomes.

## VI. The Relationship of Genes and Chromosomes

The second decade of the twentieth century is outstanding for the rapid advances made in the understanding of basic genetic principles, which were made by T. H. Morgan and his collaborators in the United States.

Morgan's interest in genetics, indeed, arose out of his studies of the sex-determining mechanism in aphids. In 1903, in accordance with the prevailing view of the time, he began experiments trying to influence the sex of these insects by varying conditions of food and temperature. In place of the traditional plants and animals hitherto used by breeders, Morgan's school subsequently adopted the fruit fly, *Drosophila melanogaster,* of which many thousands of individuals could be bred simultaneously in a single room and twenty-five generations could be obtained in a year. Notwithstanding its small size, the fruit fly shows a wealth of morphological detail, and over the years a large number of mutations arose; these mutants differed from the normal, or wild-type, flies by such characters as the color of the eyes and of the body or the shape of the wings.

As early as 1903, Walter Sutton, a pupil of both McClung and E. B. Wilson, had formulated the basic problem of genetics as follows: "We have seen reason . . . to believe that there is a definite relation between chromosomes and allelomorphs or unit characters but we have not before inquired whether an entire chromosome or only a part of one is to be regarded as the basis of a single allelomorph. The answer must unquestionably be in favor of the latter possibility, for otherwise the number of distinct characters possessed by an individual could not exceed the number of chromosomes in the germ products; which is undoubtedly contrary to fact. We must, therefore, assume that some chromosomes at least are related to a number of different allelomorphs. If then, the chromosomes permanently retain their identity, it follows that all the allelomorphs represented by any one chromosome must be inherited together."

The concept of a chromosome as a system of linked genes was experimentally verified by Morgan and his colleagues. In *Drosophila melanogaster*, the cytological evidence showed the presence of four pairs of chromosomes in each dividing cell, and, as a result of breeding experiments, four independent linkage groups were constructed (Bridges, 1914; Metz, 1914). Moreover, it was established that each individual linkage group is far from being a random assemblage of genes. Although linked genes tend to stay together during gamete formation, they may occasionally be separated through crossing-over. The more closely two genes are situated together on the same chromosome, the less likely is the chance that they will be separated by crossing-over. In this way information can be obtained about the distance separating any two pairs of genes and by combining the data from different gene pairs it is possible to construct linkage maps, which show the positions of the genes along a chromosome.

The physical basis of crossing-over was first formulated by Janssens (1909) in his "chiasmatype" theory. This involves the breaking and reunion of two out of four chromatids following chromosome duplication. That the chiasmata, which are visible during prophase of the first meiotic division, are related to the process of crossing-over has been proven beyond doubt. Nevertheless, the details of the processes involved are still the subject of active discussion (see Whitehouse, 1969).

Linkage and crossing-over have been observed in a vast variety

of organisms among plants and animals, including man, as well
as in microorganisms. Indeed, the concept of a chromosome bearing
a system of linked genes, in which each gene occupies a definite
position, has become the foundation upon which genetic theory
has been built. Therefore, it is not surprising that the assumption
is often made that a chromosome is nothing but a system of linked
genes, but this corollary is not justified either by the rules of logic
or by the scientific evidence.

On the subject of sex determination, the idea that chromosomes
are nothing but a collection of genes runs into serious obstacles,
and it is largely due to the heroic efforts of Bridges in circumnavi-
gating the factual obstacles that the theory appeared to have been
saved. Part of the conflict is illustrated in the following quotation
by Bridges (1925). "In dealing with sex and its determination,
attention has been most sharply focused upon forms with separate
sexes and upon the visible difference between the chromosome
groups of the two sexes. The result has been that the formulation
of sex determination has remained in terms of chromosomes, while
the modern unit of determination is the gene; and also the subject
of sex has been rather separated off from the main body of heredity.
My discussion will be largely a process of resolving chromosomes
into component genes, and showing that the conception of the
nature and action of genes as gained from the study of nonsexual
characters is valid in interpreting sex phenomena."

In order to understand the problem posed by the undoubted
presence of sex chromosomes and the failure to isolate sex-determining genes, the classical findings on sex determination in *Drosophila* will be briefly reviewed.

## VII. Sex Determination in *Drosophila*

In *Drosophila melanogaster*, the mitotic chromosome number
is eight. The sex chromosome of the female consists of a pair of
chromosomes, in which the centromere is situated near one end
of the chromosome. Such chromosomes are now referred to as
"acrocentric," whereas in classical preparations they appeared to
be rod-shaped (Bridges, 1916). The Y chromosome is somewhat

larger than the X, and its centromere divides the chromosome into two arms of unequal length. Such chromosomes are often referred to as "submetacentric," while the term "metacentric" is used when the two arms appear to be of more or less equal lengths. In the older literature, these chromosomes were described as J- or V-shaped. In the past, somatic chromosomes of *Drosophila* were not regarded as favorable material for cytological investigations, and it is gratifying that as a consequence of the advances made in human cytogenetics, the chromosomes of *Drosophila* have also benefited; excellent preparations have recently been obtained by Barigozzi *et al.* (1966) (Fig. 1.3).

The main body of work carried out by Bridges on sex determination was based on the use of genetic markers, although the results were verified by chromosome analysis. The X chromosome in *Drosophila* carries a large number of genes. The gene *vermilion* brightens the eye color when compared with the normal red. Being recessive, the gene will show its effect only if present in two doses in females, but a single dose is sufficient in males, since only one X chromosome is present. This accounts for the characteristic pattern of inheritance of sex-linked characters, i.e., characters which are determined by genes on the X chromosome. A female fly with vermilion eyes mated to a male with normal red eyes will have male offspring with vermilion eyes and female offspring with red eyes (Fig. 1.4). However, this striking pattern is broken occasionally, when a red-eyed male or a vermilion-eyed female may emerge from such a mating. Bridges (1916) assumed that the exceptional flies were the results of nondisjunction of the two X chromosomes during the process of egg formation (Fig. 1.5). If an egg containing two X chromosomes is fertilized by a sperm bearing a Y chromosome, the result would be á female, both of whose X chromosomes are derived from her mother and who would, therefore, show her mother's sex-linked characters. On the other hand, if an egg lacking an X chromosome is fertilized by an X-bearing sperm, the resultant fly would be a male showing the sex-linked characters of his father. The chromosome constitutions of the XO males and the XXY females were confirmed by analyzing mitotic divisions in spermatogonia and oogonia. The existence of flies with three X chromosomes was first postulated on theoretical grounds, but such flies were

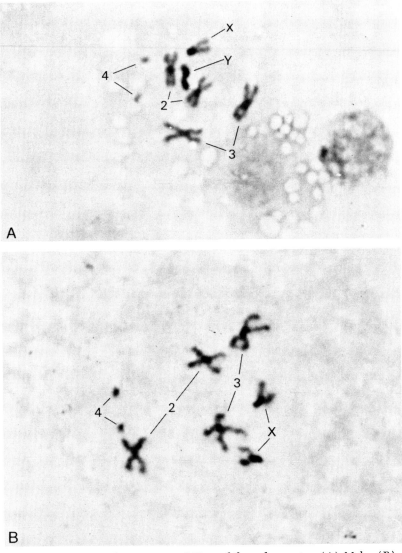

*Fig. 1.3.* Somatic chromosomes of *Drosophila melanogaster.* (A) Male; (B) female. Note heteropycnotic staining of a large part of the X chromosome, of the entire Y chromosome, and the centromeric regions of chromosomes 2 and 3. Acetic orcein stain. (Photographs contributed by C. Barigozzi.)

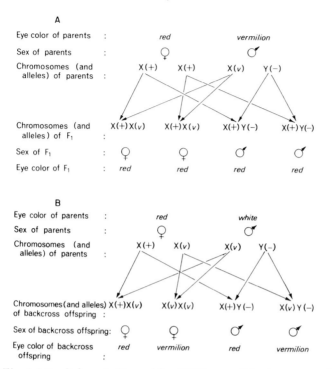

**Fig. 1.4.** Sex linkage in *Drosophila*. (Wild-type allele shown as + .)

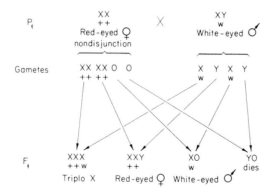

**Fig. 1.5.** Nondisjunction in *Drosophila* female.

subsequently found among the offspring of triploid females (Bridges, 1921a). They proved to be females with a high death rate; the survivors were sterile. Flies completely lacking an X chromosome have never been found.

The finding that XO flies were male, while XXY flies were female, ruled out the possibility that the Y chromosome in *Drosophila* might have a male-determining function. Although XO males appeared normal, they proved to be sterile. Stern (1929) investigated Y chromosomes with various deletions and found that in order to ensure fertility, regions on both the long and the short arm of the Y chromosome had to be present. These regions have been called "fertility factors."

As a result of mating a female which proved to be triploid, Bridges (1921a) obtained 96 females, 9 males, and 80 flies which were intermediate between males and females. The external genitalia were predominantly female. The gonads were typically rudimentary ovaries, although sometimes one was an ovary and one a testis, or an ovotestis; spermathecae (a female characteristic) as well as sex combs (a male characteristic) were present. While the abdomens were intermediate between male and female, there was a good deal of variation regarding sexual phenotype. All intersexes were sterile. They were also large and coarse bristled. Chromosome investigation showed that the two large autosomes were present in triplicate but that there were only two X chromosomes (the small dotlike chromosome was present either in triplicate or in duplicate).

The discovery that flies having two X chromosomes and three sets of autosomes were intersexes made it clear that femaleness was due not simply to the number of X chromosomes, but rather to the ratio of the number of X chromosomes to that of autosomes. Bridges explained this finding by postulating that there are genes on the X chromosome which have a tendency to produce female characteristics and that there are genes on the autosomes which produce the male characters. In this way, the causes underlying sex determination became incorporated in the theory of genic balance. Before discussing this theory in more detail, brief mention will be made of the findings on sex determination in the moth, *Lymantria*.

## VIII. Intersexes in *Lymantria*

A vast amount of data on the gypsy moth, *Lymantria dispar,* were accumulated by Goldschmidt (1934) for the specific purpose of explaining sex determination in terms which go beyond a formal chromosomal mechanism.

The gypsy moth belongs to the insect order Lepidoptera, in which the female in heterogametic (this chapter, Section V). It was chosen because intersexes are relatively common in this species. Goldschmidt distinguished between gynandromorphs (which are mosaics of male and female parts and contained cells with both male and female sex chromosome constitutions) and intersexes (or individuals with abnormal sex development, in which all the cells have the same sex chromosome constitution). This distinction is useful in insects, although it is not necessarily applicable to other organisms.

In *Lymantria,* Goldschmidt found that intersexes occurred regularly among the offspring of crosses between different geographical races. For instance, crosses between Japanese females and European males resulted in normal male and female offspring, but the reciprocal crosses produced normal males and intersexual females. Also, when the offspring of the first cross were mated together, the female offspring were normal but up to one-half of the males were intersexual.

These findings were explained on the assumptions that (1) different races of *Lymantria* differ in the strength of their sex determinants and that in the present example the Japanese race is a strong sex determiner and the European race a weak one and that (2) the sex of a moth is determined by the relative strength, or balance, of a female-determining factor, which is inherited from the mother, and male-determining factors, which are located on the X chromosome. Regarding the location of the female-determining factor, Goldschmidt (1934) at different times favored the view that it was carried either on the Y chromosome or in the cytoplasm. The subject generated some controversy. Winge (1937) put forward the view that there are female-determining genes on the X chromosome, male-determining on the Y chromosome, and, in addition, there are sex-determining genes on the autosomes.

It is clear that the exact roles of the X and the Y chromosomes in sex determination can be established only on the basis of reliable data regarding the sex of individuals with abnormal sex chromosome constitutions, particularly with XXY and XO sex chromosomes, and such data are not available in *Lymantria*. However, the *Lymantria* experiments provide a classical example of the relationship of sex differentiation to rates of growth. To begin with, the time taken for the development of different larval stages differ in the two sexes and also in different geographical strains; the growth of the gonads, however, does not necessarily follow that of the rest of the body. In young larvae, the testes grow much faster than the ovaries, whereas the main growth of the ovaries occurs after pupation. Since different races of *Lymantria* differ both in the general timing of development and in the growth and differentiation of the gonads, it is not surprising that racial crosses result in disharmonious growth leading to the formation of intersexual gonads (Goldschmidt, 1934).

Differences in rates of growth come within the province of quantitative genetics, which postulates large numbers of genes underlying each measurable character (Chapter 2). The theory of genic balance is related to this concept and will be discussed first, since it marks the culmination of the classical theory of the genetics of sex determination.

## IX. The Theory of Genic Balance

The theory of genic balance originated in the observation that *Drosophila* flies which lacked one chromosome of the small fourth pair, although reasonably viable, showed certain phenotypic differences compared with normal flies (Bridges, 1921b). Subsequently, this theory was elaborated (Bridges, 1925, 1939) to fit the facts of sex determination.

The results of linkage experiments seemed to indicate that genes affecting any particular organ or character were distributed more or less at random along the different chromosomes. For instance, a similar ruby eye color was found to be produced by five mutants on the X chromosome and by several others on the second and third chromosomes.

The central idea of the theory of genic balance is that each character of the adult is produced by the joint action of all the genes on the entire complement of chromosomes. Some of these genes tend to drive the development in a particular direction or to increase it (so-called "plus genes"); others tend to drive development in the opposite direction or to decrease it (so-called "minus genes"). Each of these genes is a producer in this joint effect, although for each particular feature some genes are more effective than others. The stage of development shown by each feature of a normal adult corresponds to a net effect of these opposed tendencies or a point of balance between these plus and minus genes. It follows that a change in the end point or balance may be achieved not only by a gene mutation, but also by the loss, or the addition, of an entire chromosome, or part of a chromosome, provided this results in a different proportion of plus and minus genes.

Bridges (1922) himself recognized the similarity of an abnormal autosome number and the sex chromosome dimorphism. Flies with only one fourth chromosome show a change in many characters, including smaller size, smaller bristles, later hatching, impaired viability, and darker trident pattern. The explanation given was that the characters which were less strongly developed were affected by plus genes on the fourth chromosome, which were in short supply because of the lack of one of these chromosomes, while the more strongly developed characters, such as darker trident pattern, were caused by minus genes on the fourth chromosomes. Turning to the X chromosome, the gene *eosin* causes males to have a paler eye color than females, the interpretation being that the X chromosome contained an excess of minus genes affecting this character. Bridges thought that differences in sex organs could be explained in the same way.

The idea that one or more genes on the X chromosome were responsible for sex determination had its origin in the early years of *Drosophila* genetics. Indeed, the term "sex linkage" bears witness, as is clear from the following quotation from Morgan and Bridges (1916): "This term (sex-linked) is intended to mean that such characters are carried by the X chromosomes. It has been objected that this use of the term implies a knowledge of a factor for sex in the X chromosome to which the other factors in that chromosome are linked, but in fact we have as much knowledge in regard to the occurrence of a sex factor or sex factors in the X chromosome as we have for other factors." The uncertainty of whether one or more sex factors were present was thought to be due to the fact that in *Drosophila* there is no crossing-over in the male. This argument became largely invalidated with the discovery that the Y chromosome in *Drosophila melanogaster* is not male-determining.

The task of localizing female determining genes on the X chromosome was resumed by Dobzhansky and Schultz (1934), following Muller's (1930) discovery that chromosomes could be broken by X rays. The experiments involved irradiating flies and mating them to others with specially marked X chromosomes. The best test material proved to be triploid intersexes, because their sexual development can be altered with relative ease. It was found that by adding pieces of X chromosome to triploid flies with two complete X chromosomes, the development of the intersexes was shifted in a female direction. This effect was obtained by all parts of the X chromosome, apart from the inert region (Chapter 4, Section III). In general, the degree of feminization achieved was proportional to the length of the added piece of chromosome. If this was sufficiently long, the intersexes were turned into fertile females, and there was no particular part of the X chromosome which was essential for this purpose. These results seemed to rule out the existence of a single sex-differentiator gene and were interpreted as indicating the presence of large numbers of female-determining genes situated along the length of the X chromosome.

Of the many genes involved, those which affected normal sex development were called "sex producers" (Bridges, 1939). Among these were included the so-called "sex modifiers" whose affect on sexual development was relatively slight. A sex producer could become a "sex differentiator" by mutation. Some of these were thought to show up only in special conditions such as the intersexes with triploid autosomes and two X chromosomes.

In order to illustrate the difference between a sex producer and a sex differentiator, Bridges used the analogy of a balance. "As equal weights are added to the pan of a balance, the beam finally tips. Let us say ten weights left it untipped and after the eleventh was added it tipped. But this eleventh weight has no more intrinsic significance in the tipping than each of the weights added before. If any one of them had been omitted the beam would not have tipped when this one was added. All together produce the result. The last one is the differentiator of the position of the beam."

In this scheme, therefore, both the sex producers and the sex differentiators of Correns were assigned to genes, although there appears to be some confusion whether the sex differentiator acts by the specific effect of a mutant or in a quantitative way by the number of equal genes present. The quantitative aspect of the concept was in accordance with the ideas put forward by Goldschmidt (1920) with regard to sex determination.

Since sex in *Drosophila melanogaster* is determined by the ratio of the number of X chromosomes to sets of autosomes, Bridges postulated that the net male tendency of a set of autosomes is less than the net female tendency of an X. By assigning a value of 100 to an X chromosome and a value of 80 to haploid set of autosomes, the "sex indexes" could be neatly arranged, as shown in Table 1.1; and so the problem of sex determination was reduced to an arbitrary numerology. The concept of the supersexes was, of course, entirely theoretical and bore no relation to the

TABLE 1.1
SEX PHENOTYPE AND X/AUTOSOMAL RATIO IN
*Drosophila melanogaster*[a]

| Sex type | No. of X chromosomes (X) | No. of sets of autosomes (A) | Sex index (100X/ 80A) |
|---|---|---|---|
| Superfemale | 3 | 2 | 1.88 |
| Female |  |  |  |
| 4N | 4 | 4 | 1.25 |
| 3N | 3 | 3 | 1.25 |
| 2N | 2 | 2 | 1.25 |
| N | 1 | 1 | 1.25 |
| Intersex | 2 | 3 | 0.83 |
| Male | 1 | 2 | 0.63 |
| Supermale | 1 | 3 | 0.42 |

[a] After Bridges, 1939.

anatomy or physiology of the flies with abnormal numbers of X chromosomes. On a more fundamental level, the nature and function of the male- and female-determining genes was entirely unknown. At a time when these questions were not at the forefront of genetical inquiry, Bridges scheme provided a reasonably consistent model of sex determination in *Drosophila* and was widely adopted.

The theory of genic balance is still widely invoked today (see Stern, 1967). It is used to explain abnormalities of development resulting from abnormal chromosome constitutions, although its power of predicting the type of abnormality resulting from any particular chromosomal change is extremely low. The theory is currently meeting a specific challenge through the findings of triploidy in man. In *Drosophila* and many species of plants, individuals with three instead of two sets of chromosomes are viable and not strikingly different from the normal, and it has generally been accepted that this is due to the fact that the balance of genes is not disturbed. In man, however, the condition is highly lethal and triploidy is one of the most common chromosomal abnormalities found in spontaneously aborted embryos (Chapter 6). Although the cause for this lethality is not yet proved beyond doubt, it is becoming increasingly likely that in the future the effects of different chromosome constitutions on development will have to be sought in mechanisms which do not require the existence of specific hypothetical genes.

There is little doubt that the difficulties encountered in the task of formulating a genetic theory of sex determination were a disappointment to classical geneticists. "It has sometimes been felt that the determination of sex offered the best opportunity for the study of the manner of action of genes . . . , it now appears, however, that it will be more profitable to study simpler situations, and it is to these that attention is now more often directed" (Sturtevant, 1965). The "simpler situations" come within the province of biochemical genetics, and the results of these studies have led to our present knowledge of the nature and function of genes (Chapter 3). The results of classical studies, however, have made it clear that sex determination is closely connected with quantitative genetics, or, more precisely, with genetics of continuous variation; this will now be considered.

# Chapter 2

## Genetics of Continuous Variation

### I. Introduction

The characters most favored by geneticists differ from each in a clear-cut and unambiguous manner, but these do not, of course, exhaust the variation observed between members of the same species. The stature of human beings varies continuously from the tallest to the shortest; to analyze this type of variation, simple classification must give place to measurements.

The foundations underlying the genetics of continuous variation have recently been reviewed by Sewall Wright (1968). The present chapter will be confined to a few points which appear to be of special relevance to genetic theory as a whole and have a particular bearing on the problem of sex differentiation.

The size obtained by an individual, human as well as animal or plant, is partly due to heredity but is also affected by environmental factors such as nutrition. This was undoubtedly known long before the advent of modern science.

When large numbers of living things are measured, the results tend to fall into characteristic patterns. More than 100 years ago, the Belgian astronomer and statistician, Quetelet, arranged the measurements of 26,000 soldiers in the United States army and found that the measurements were distributed around the mean, or average, in a symmetrical fashion: the further away from the mean, the fewer the numbers of individuals (Table 2.1); moreover, the distribution of the numbers of individuals per class gave a good approximation to the binomial formula and fitted a normal curve.

When one compares the heights of parents and children, new patterns become apparent, as was clearly recognized by Francis Galton. In his book "Natural Inheritance," published in 1889, Galton wrote: "One of the problems to be dealt with refer to the curious regularity commonly observed in the statistical peculiarities of great populations during a long series of generations. The large do not always beget the large, nor the small the small, and yet the observed proportion between the large and the small in each degree of size and in every quality, hardly varies from one generation to another."

TABLE 2.1
STATURE OF 25,878 UNITED STATES SOLDIERS[a]

| Height of soldiers (mm[b]) | No. of soldiers measured to the nearest inch | No. per 1000 | |
|---|---|---|---|
| | | Observed | Calculated |
| 1397–1524 | 31 | 1 | 2 |
| 1525–1549 | 15 | 1 | 3 |
| 1550–1575 | 50 | 2 | 9 |
| 1576–1600 | 526 | 20 | 21 |
| 1601–1626 | 1237 | 48 | 42 |
| 1627–1651 | 1947 | 75 | 72 |
| 1652–1676 | 3019 | 117 | 107 |
| 1677–1702 | 3475 | 134 | 137 |
| 1703–1727 | 4054 | 157 | 153 |
| 1728–1753 | 3631 | 140 | 146 |
| 1754–1778 | 3133 | 121 | 121 |
| 1779–1803 | 2075 | 80 | 86 |
| 1804–1829 | 1485 | 57 | 53 |
| 1830–1854 | 680 | 26 | 28 |
| 1855–1880 | 343 | 13 | 13 |
| 1881–1905 | 118 | 5 | 5 |
| 1906–1930 | 42 | 2 | 2 |
| 1931–2007 | 17 | 1 | 0 |

[a] From Quetelet, 1871.
[b] Presumably converted from inches.

Galton measured sweet peas and moths, but particularly human beings, comparing the stature of children with that of their parents as well as measuring the variability of the children within a sibship. It is of interest to note that he dealt with the observed sex difference in stature by multiplying all heights of females by a factor of 1.08. By tabulating the average heights of the parents with that of their children (Table 2.2), Galton developed the concept of correlation, although apparently it had been discovered before him. It will be seen from the table that parents, whose height is either above or below the mean, are likely to have children whose height deviates from the mean in the same direction, but to a lesser extent.

The mathematical concepts developed by Galton were elaborated and improved by Karl Pearson (1904). Pearson was a mathematician who was preoccupied with mathematical and statistical descriptions of quantitative inheritance. He was extremely skeptical of Mendelian explanations. Thus, he became allied with Weldon, a mathematically inclined zoologist, against Bateson. The result was that Mendelian genetics and statistics began as entirely separate disciplines, their only point of contact being frequent polemics between the biometrical and the Mendelian schools. It is often thought that this state of affairs impeded the early progress of genetics, but it is at least equally likely that the continued attacks by Pearson and his biometric allies acted as an impetus to Bateson and his colleagues to overcome the difficulties posed by some genetical data and develop and strengthen the foundations of Mendelian theory.

Very soon after the rediscovery of Mendel's papers, Bateson and Saunders (1902) wrote that a character, such as stature, must certainly be determined by more than one pair of allelomorphs. Even if only four or five pairs were involved, the various combinations might give so near an approach to a continuous curve that the purity of the elements would be unsuspected, and their detection practically impossible. Although a discontinuous character, the shape of the comb in fowls was shown to be affected by two pairs of alleles acting in unison (Bateson et al., 1908). Different breeds of poultry differ in the shape of their combs. When fowls with so-called "rose" combs were crossed fowls having "pea" combs,

TABLE 2.2
HEIGHTS OF PARENTS AND ADULT CHILDREN[a]

| Heights of mid-parents[b] (inches) | No. of adult children with heights of (in inches) | | | | | | | | | | | | | | Total number | |
|---|---|---|---|---|---|---|---|---|---|---|---|---|---|---|---|---|
| | >62.2 | 62.2 | 63.2 | 64.2 | 65.2 | 66.2 | 67.2 | 68.2 | 69.2 | 70.2 | 71.2 | 72.2 | 73.2 | >73.2 | Children | Mid-parents |
| 72.5 | — | — | — | — | — | — | — | 1 | 2 | 1 | 2 | 7 | 2 | 4 | 19 | 6 |
| 71.5 | — | — | — | — | 1 | 3 | 4 | 3 | 5 | 10 | 4 | 9 | 2 | 2 | 43 | 11 |
| 70.5 | 1 | — | 1 | — | 1 | 1 | 3 | 12 | 18 | 14 | 7 | 4 | 3 | 3 | 68 | 22 |
| 69.5 | — | — | 1 | 16 | 4 | 17 | 27 | 20 | 33 | 25 | 20 | 11 | 4 | 5 | 183 | 41 |
| 68.5 | 1 | — | 7 | 11 | 16 | 25 | 31 | 34 | 48 | 21 | 18 | 4 | 3 | — | 219 | 49 |
| 67.5 | — | 3 | 5 | 14 | 15 | 36 | 38 | 28 | 38 | 19 | 11 | 4 | — | — | 211 | 33 |
| 66.5 | — | 3 | 3 | 5 | 2 | 17 | 17 | 14 | 13 | 4 | — | — | — | — | 18 | 20 |
| 65.5 | 1 | — | 9 | 5 | 7 | 11 | 11 | 7 | 7 | 5 | 2 | 1 | — | — | 66 | 12 |
| 64.5 | 1 | 1 | 4 | 4 | 1 | 5 | 5 | — | 2 | — | — | — | — | — | 23 | 5 |
| Below 64.5 | 1 | — | 2 | 4 | 1 | 2 | 2 | 1 | 1 | — | — | — | — | — | 14 | 1 |
| Total | 5 | 7 | 32 | 59 | 48 | 117 | 138 | 120 | 167 | 99 | 64 | 40 | 14 | 14 | 924 | 200 |

[a] After Galton, 1889.
[b] Mid-parent expresses an ideal person of composite sex, whose stature is half-way between the stature of the father and the transmuted stature of the mother (i.e., mother's stature × 1.08).

the first generation ($F_1$) offspring had "walnut" combs. In the second ($F_2$) generation yet another type of comb, known as "single," appeared. The numbers of different combs, which at first seemed difficult to interpret, were found to show excellent agreement with the ratio 9 walnut: 3 pea: 3 rose: 1 single. Such a ratio would be expected on the assumption that the walnut requires the presence of two separate dominant genes, the pea and rose each require one dominant gene, while the single is a double recessive. The validity of this assumption can, of course, be simply verified by appropriate crosses. Examples of other characters which were determined by two or more pairs of alleles soon became apparent, for instance, pericarp color in maize (*Zea mays*) (East, 1910), and color as well as more obviously quantitative characters in oats and wheat (Nilsson-Ehle, 1909). Even if only a few pairs of genes determine the same character, quite large numbers of different combinations in the expression of the character would result. If there were $n$ genes with dominant effect, the number would be $2^n$; if there were no dominance, the number would be $3^n$. Nilsson-Ehle (1909) suggested that the inheritance of quantitative characters might be explained by the action of several, or many, pairs of genes.

Another important fact concerning the inheritance of quantitative characters emerged at that time. When pure bred parents, which differ in height or weight are crossed, the first hybrid ($F_1$) generation tends to be intermediate between the parents and fairly uniform in size. When these hybrids are crossed among themselves, the $F_2$ generation are much more variable, and it is possible, by selection, and in later generations, to produce types which are more extreme than either parent. This was established both in plants (Emerson and East, 1913—length of cob and other quantitative characters in maize) and in animals (Punnett and Bailey, 1914—weight of fowl). The greater variance of the $F_2$ hybrids compared with their $F_1$ parents argues against the concept of blending inheritance and in favor of the idea of segregation and recombination of large numbers of genes in the genetically uniform but heterozygous $F_1$ generation.

The ultimate union of Mendelian genetics with statistics was to a large measure due to the work of the great Danish biologist,

Johannsen. He combined biological insight with mastery of statistical techniques and added to this clarity of exposition. His "Elemente der exakten Erblichkeitslehre," published in 1909, remains illuminating to this day.

Johannsen was the originator of the terms "gene," "phenotype," which he defined as "apparent type" (*Erscheinungstypus*), and "genotype" ("inherent type" or *Anlagetypus*). He was careful to point out, however, that the term "genotype" was of no practical use, since it was never known in its totality; he commended to use only "genotypic difference." In addition, he was the originator of the concept of the pure line. This comprises all individuals which have arisen from a single organism, which is habitually self-fertilized and, therefore, entirely homozygous. The bean, *Phaseolus vulgaris*, is an example of a plant which reproduces by self-fertilization, and in this species, Johannsen established 19 pure lines. The weights of beans varied in different lines and also within the same pure line, but when the weights of beans of parents and offspring were compared, it was found that only differences between lines were inherited; within a pure line, the weight of the mother bean had no effect on the weights of the daughter beans. It thus became established that the continuous type of variation is divisible into two components: one which is inherited and must, therefore, be determined by the genetic material, and the other which is not inherited and is generally referred to as "environmental," that is caused by agencies outside the genetic material.

If is, of course, the inherited part which is of most concern to geneticists and work has been concentrated in two directions. One is the elaboration of statistical techniques with which the data of continuous variation and its inheritance can be handled, and the other is an attempt to understand the basic realities underlying the genetics of continuous variation.

Once it is realized that a part of the continuous variation exhibited by living organisms is inherited and also that these characters do not seem to obey the rules of Mendelian genetics, the question arises whether they are determined by Mendelian genes or by some other part of the genetic material. Most geneticists seem to favor the first alternative and to some, indeed, this has crystallized into a dogma. Thus Falconer (1964) in "Introduction to Quantitative Genetics" writes: "It is, nevertheless, a basic premise of quantitative genetics that the inheritance of quantitative differences depends on genes subject to the same laws of transmission and having the same general properties as the genes whose transmission and properties are displayed by qualitative differences." One of the principal founders of this view was R. A. Fisher who, in a classical paper published in 1918, pointed out that the observed differences in the stature of brothers are too large to be ascribed entirely to the effects of the environment. The simplest hypothesis is that characteristics such as stature are determined by a large number of Mendelian factors with cumulative effects and that the large variance among children of the same parents is due to the segregation of those factors in respect to which the parents are heterozygous. The hypothesis of multiple genes with additive effects also fits the increased variance in the $F_2$ generation as compared with their $F_1$ parents. The theory clearly showed excellent agreement with observed quantitative data, even though it lacked any concrete biological basis. Haldane (1932) wrote as follows: "No procedure of this kind [i.e., detecting individual genes] is as yet possible in the case of such an apparently continuously variable character as the height of men or weight of rabbits. Ultimately it may well be found that of the genes influencing human height some act through the thyroid gland, others through the pituitary, others through the gonads in delaying maturity, others again more directly on the bones, and so on. That is mere speculation. At present we cannot even prove conclusively that such continuously varying characters are due to genes at all. But we can render it extremely plausible."

At a time when nothing was known about the nature of genes or their mode of action it was clearly impossible to arrive at any certain conclusion about the hereditary basis of continuous variation. For practical purposes it seemed logical to many geneticists to divide the genes into two classes. Those with large effects which could be individually recognized, obeyed the rules of Mendelian genetics, and gave rise to discontinuous variation, and genes whose effects were too small to be individually recognized but which, allied with genes with similar effects, gave rise to continuous variation. The question whether the two types of variation were caused by the same type of gene or whether there were two classes of genes which were basically dissimilar gave rise to some controversy.

## II. Major Genes with Minor Effects

Mendelian genes with discontinuous effect may, in addition, show subsidiary effects which may fall into the class of continuous variations. This is well illustrated by the findings in phenylketonuria, described by Penrose (1951). Phenylketonuria is a disease which is caused by a Mendelian gene. The most important clinical effect is mental retardation, usually of a severe grade. The primary effect of the gene is a defect in the enzyme phenylalanine 4-hydroxylase, which normally converts phenylalanine into tyrosine (see Harris, 1970). Patients having the recessive gene in homozygous dose have a highly increased level of phenylalanine in their blood and, although there is a certain amount of variation both in normal people and in phenylketonurics, the two distributions are entirely separate (Fig. 2.1). When intelligence quotients are measured, a small amount of overlap between phenylketonurics and nonphenylketonurics is found. In addition to these major effects, the gene has a much smaller effect in reducing the size of the head and in lightening the color of the hair. In both these cases there is considerable overlap between phenylketonurics and nonphenylketonurics and neither characteristic would be suitable for classifying the presence or absence of the disease.

Not only abnormal genes, but normal gene variants may have quantitative effects in addition to the more obvious qualitative differences which they produce. The gene locus for red cell acid phosphatase has at least three alleles, which produce structurally different enzymes with different electrophoretic mobilities (Harris, 1970). When the activity of the enzyme is measured quantitatively, it is found that the mean activity of the B variant is about 50% higher than that of the A variant, while the activity of the C variant is somewhat higher still. Harris suggested that the observed differences in activity of other enzymes might similarly be related to different structural variants of the enzymes.

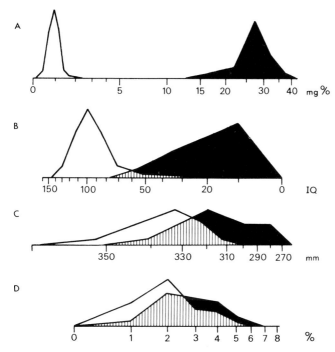

**Fig. 2.1.** Frequency distributions of phenylketonurics (right) compared with control population (left). (A) Phenylalanine in blood plasma (mg%); (B) intelligence (Binet, I.Q.); (C) head size (length plus breadth in mm, corrected for sex); (D) hair color (reflectance % at 100 nm, corrected for age). The areas of the distributions are proportional to their respective standard deviations. Black: distribution of measurements of phenylketonurics; white distribution of measurements of controls; shaded: distribution of measurements common to both populations. (From Penrose, 1951.)

Grüneberg (1952, 1955) made the more general suggestion that direct gene effects would give rise to discontinuous variation, while the more remote effects of genes results in continuous variation and that, moreover, the basis of quantitative genetics is to be sought in the distant effect of Mendelian genes.

Genes which show more than one observable effect are said to be "pleiotropic." This term was originally introduced by L. Plate (1910) to denote genes which produce multiple changes during the development of an organism. It has subsequently been shown, particularly by Grüneberg (1938, 1943), that where a gene appears to have a number of different effects, these can sometimes be traced back to a single change produced by the gene. In this way, an entire syndrome can be shown to be related by a "pedigree of causes" to a single gene effect, and pleiotropism can thus be explained as a complex effect of one changed reaction during the development of the organism.

Clearly, some quantitative variation can be confidentially regarded as resulting from changes in Mendelian genes. However, to conclude that all hereditary quantitative variation can be accounted for in this way is clearly unjustified. Such a conclusion could be arrived at on *a priori* grounds only if it could be shown that the entire genetic material is taken up by Mendelian genes, an assumption which we now know not to be true (Chapter 4, Section VI). Since the argument cannot be settled by theoretical reasoning, it will become necessary to examine the biological basis of quantitative variation in specific systems. This topic will be pursued further in a later chapter.

### III. Quasicontinuous Variation

The topic to be discussed in this section is to some extent the reverse of the subject described in the previous section. Quantitative variation—we shall leave aside its mode of origin for the time being—can in certain circumstances give rise to a type of variation which appears in practice to be discontinuous. This process may be of special relevance in the development of sex differentiation of higher organisms.

Guinea pigs (*Cavia porcellus*) normally have four digits on the front feet and three digits on the hind feet, but an extra small digit occurs not uncommonly. The incidence of an extra toe varies in different strains and it can be increased by selection. Therefore, although the condition has a genetic basis, it cannot be due to the action of a single gene. Wright (1934a,b) suggested that many genes are involved and that the presence of an extra toe is the result of a physiological threshold being reached.

Grüneberg (1952, 1964) concluded that certain skeletal variants in the mouse (*Mus musculus*) have a similar genetic basis. For instance, the third molar tooth is almost invariably present in mice of the $C_{57}$ Black strain but is absent in nearly one-fifth of the mice of the CBA strain. When the two strains are crossed, the tooth is present in the $F_1$ hybrids as well as in the $F_2$ generation. The differences are correlated with differences in size of the molars. Whereas in $C_{57}$ Black, the third molars are fairly large and of uniform size, those in the CBA strain are smaller and more variable. The sizes of molars in the $F_1$ and $F_2$ generations approach those of the $C_{57}$ Black. Grüneberg concluded that the absence of the third molar is the result of its small size. When a tooth rudiment fails to reach a critical size by a certain time of development, it is unlikely to develop into a tooth.

The absence of the third molar is an example of a qualitative character with a quantitative basis. Grüneberg (1952) has called the process "quasicontinuous variation," while the name given to it by Wright (1968) is "threshold dichotomy," which is defined as "two sharply alternative types (which) clearly depend on whether the physiological state at a critical time, resulting from a complex of interacting genetic and environmental factors, is on one side or the other of a certain threshold."

In man, there are a number of serious developmental abnormalities, which have a genetic basis but are not transmitted in a clearcut Mendelian fashion. Anencephaly (failure of brain development), spina bifida (nonclosure of posterior part of neural plate), and cleft palate are examples. The possibility that these conditions are examples of quasicontinuous variation has been discussed (Penrose, 1957; Edwards, 1960; Fraser and Pashayan, 1970). Carter *et al.* (1968) suggest that the most plausible explanation to account for these conditions is "polygenic inheritance," with much interaction with "intrauterine environmental variables," although, since one classifies only two states, normal versus abnormal, a threshold mechanism must be superimposed.

Wright (1968) cites as an example of threshold dichotomy those lower organisms which have sharply defined sexes but no sex chromosome mechanism. The marine worm, *Bonellia viridis*, is a well-known example of an animal in which sex is determined to a large extent, if not entirely, by environmental agents; isolated larvae develop into females, while larvae settling near a female develop into males (Baltzer, 1914; Bacci, 1965).

The possibility will be discussed below (Chapter 5) that in higher organisms, in which sex chromosomes are present, the process of sex differentiation may also be due to a threshold dichotomy and that the sex chromosomes may be responsible both for the continuous variation and for the threshold effect.

## IV. Polygenes

In the controversy regarding the nature of the genetic material which is responsible for continuous variation Mather (Mather and Harrison, 1949; Mather and Jinks, 1971) has taken the stand that the units involved may be different from those giving rise to Mendelian differences. Based on the assumption that continuously variable characters are caused by multiple genes, Mather set out to show that the assembly of multiple genes, which he called "polygenes" [the adjective "polygenic" had already been introduced by Plate (1932) to denote characters which were determined by many genes] exhibited one of the fundamental characteristics of ordinary genes, namely, linkage.

The number of bristles, or chaetae, on the abdomen of *Drosophila melanogaster* is a continuously varying character. The mean number of abdominal chaetae varies in different strains, and there is also a sex difference within the same strain, females having a greater number than males. By using major genes as markers for the two large autosomes and the X chromosome (and making use of the fact that there is no recombination in *Drosophila* males), Mather was able to show that each of these chromosomes affected the number of abdominal chetae, or, in his terminology, all these chromosomes carried polygenes affecting chaeta number.

By selective breeding from flies with either high or low numbers of abdominal chaetae, it was possible to obtain a progressive increase or decrease in chaeta number in successive generations. However, as a result of the inbreeding involved, the fertility of the flies was reduced. When selection for high chaeta number had continued for 20 generations, fertility fell so alarmingly that selection had to be discontinued. When the flies were subsequently outcrossed, fertility rose and the chaeta number fell. This might have suggested a pleiotropic effect of the same gene causing both high chaeta number and low fertility, but the results of subsequent experiments disproved this. When selection was resumed, the chaeta number soon rose to its previous peak, but this time without a corresponding fall in fertility. Mather concluded that the original linkage between the polygenes controlling high chaeta number and low fertility had been broken by recombination, as a result of which the characters were separated.

While selecting for high chaeta number, other unexpected effects became apparent. Normal *Drosophila* females have two spermathecae, but an appreciable proportion of females in selected lines had abnormal numbers varying between none and five. These differences must have arisen as correlated responses to the selection for high chaeta number, and it was postulated that the chromosome segments, which affected these changes, carried polygenes of diverse systems (Mather and Harrison, 1949).

The concept of the polygene thus emerged as a chromosomal region, which affected quantitative characters. Because of their chromosomal nature, polygenes show linkage and recombination, but since their effects are less specific than those of Mendelian genes, and in any case usually slight, polygenes cannot be individually characterized. One positive characteristic of polygenes which distinguishes them from Mendelian genes was, however, recognized, namely, their tendency to be situated in heterochromatin, i.e., those parts of the chromosomes from which Mendelian genes are absent (Mather, 1944).

At the time when these concepts arose, hardly anything was known about the nature of Mendelian genes, while ideas about heterochromatin were vague and not generally accepted. During the last 20 years, knowledge about the gene has crystallized to such an extent that it has become a cornerstone in biology; while new discoveries about the behavior of heterochromatin have advanced the subject, which had been of doubtful respectability, to the forefront of cytogenetic research. The two topics will be discussed in the following two chapters.

# Chapter 3

## Aspects of the Gene

### I. Introduction

Ideas on the nature of the gene have undergone many changes since they originated at the turn of the century. An understanding of the current concept of the gene is clearly a prerequisite to a discussion on the genetics of sex differentiation, both in order to discard ideas which are remains of a bygone era of genetic thought and in order to arrive at new working hypotheses. The history of the gene concept has been described by Dunn (1965) and by Carlson (1966). Here we shall merely summarize a few points which appear to be of special relevance to the topic of sex differentiation.

### II. The Gene before 1950

Genetics deals with the transmission of biological characteristics from parents to offspring. In the earliest stage, there was no clear distinction between the characteristic and the agent responsible for its transmission. When describing the contrasting characteristics of the garden pea, *Pisum sativum*, Mendel (1866) used symbols to denote the characteristics themselves. Thus, yellow color of the embryo was denoted by A and green color, which behaves as a

*43*

recessive to yellow, as *a*. To Correns (1900), one of Mendel's rediscoverers, the distinction was clear: "For an explanation one must
assume, like Mendel, that after the union of the sexual nuclei,[*]
the potentiality ("Anlage") for the recessive characteristic, in our
case for green color, is prevented from developing by the "dominant"
characteristic, i.e., the yellow color; the embryos become yellow.
The potentiality, however, remains 'latent' and before the definitive
formation of the sexual nuclei there always occurs a complete separation of the two potentialities, so that one-half of the sexual nuclei
contains the potentiality for the recessive characteristic, i.e., green
color, while the other half contains the potentiality for the dominant characteristic, i.e., yellow color" (author's translation). However, to de Vries (1900), this distinction was less clear: "These
antagonistic characteristics remain ordinarily combined throughout
the entire vegetative life, one dominant, the other latent. But during
the generative period they disjoin. Each pollen grain and each
egg cell receives only one of the two" (author's translation).

Bateson (1909) was content to regard male and female gametes
as bearers of alternate characters. He developed the concept of
"unit characters" which segregated during the meiotic divisions
preceding gametogenesis. Castle (1903) stated that "every gamete
(egg or spermatozoon) bears the determinants of a complete set
of somatic characters," but in the same paper he also quotes the
"splitting . . . of the parental characters at gamete formation."

The gene was named, and its concept clearly stated, by Johannsen (1909): "The something in the gametes and in the zygote,
which is of essential importance for the character of the organism,
is generally given the ambiguous name "Anlagen." Many other ex-
pressions have been proposed, most of them unfortunately strictly
tied to definite hypothetical assumptions . . . . No hypothesis
about the nature of the 'something' should be formulated or supported. For this reason it seems simplest, to isolate from Darwin's
well known term [pangene[†]] the only interesting last syllable 'gene'

---

[*] Mendel, of course, does not mention nuclei but "germinal cells" and
pollen cells.

[†] The attribution of this term to Darwin appears to be erroneous. It was
introduced by de Vries (1889) and, although derived from Darwin's term
"pangenesis," its meaning was different.

in order to replace the ambiguous term 'Anlage' . . . . The word 'gene' is entirely free from any hypothesis, it merely expresses the proven fact that at least many characters of an organism are caused by special separable and therefore independent 'conditions,' 'Anlagen' which we want to call genes, and which are present in the gametes" (author's translation).

In the years which followed Johannsen's definition of the gene, it became clear, through the work of Morgan, Bridges, and Sturtevant, that genes occupied definite positions on particular chromosomes. The steps leading to this remarkable achievement have been described by a number of authors (Dunn, 1965; Sturtevant, 1965; Carlson, 1966). The following facts are of particular relevance.

1. At the beginning of the work with *Drosophila*, Morgan (1910a), although not a wholehearted believer in the chromosome theory of heredity, nevertheless, considered it very seriously. In this he was particularly motivated by the recent discoveries of the sex chromosomes.

2. The first mutant gene in *Drosophila*, white eye, was found to be located on the X chromosome and the first linkage map comprised six genes on the chromosome (Sturtevant, 1913). Soon more genes were added to this group and similar linkage maps were constructed for chromosomes 3 and 4.

As a result, genes could be defined by their positions on the chromosomes and so had become "loci," but their nature remained obscure. For the conclusion of his book "The Theory of the Gene," Morgan (1926) wrote as follows: . . . "it nevertheless is difficult to resist the fascinating assumption that the gene is constant because it represents an organic chemical entity. This is the simplest assumption one can make at present . . . and . . . it seems, at least, a good working hypothesis."

The first correct hypothesis of the way in which genes function arose from a study of the genetics of man. Garrod (1908) concluded that the inborn errors of metabolism, such as, alkaptonuria, cystinuria, and albinism, which he recognized as being due to the presence of a recessive allele in homozygous dose, were caused by metabolic blocks at intermediate stages, due to the absence in each case of a specific enzyme. However, the link between genes and enzymes were not recognized until it was rediscovered by

Beadle and Tatum (1941) on the basis of their own work on the bread mold, *Neurospora*.

The discovery that X rays have the effect of increasing the frequency of gene mutations (Muller, 1928) advanced the understanding of the nature of the gene in at least two ways. Calculations relating dose and mutation rates were used to estimate the size of the gene on the so-called "target theory" (Lea, 1955). The ultimately more far-reaching effect was in the use made of X rays in the production of mutations in microorganisms. By means of X rays, Beadle and Tatum (1941) produced a series of mutations in the fungus *Neurospora*, where it could be demonstrated that each was deficient in a single enzyme. As a consequence, Garrod's work was rediscovered and the "one gene–one enzyme" hypothesis was put on a firm experimental basis.

However, the nature of the genetic material was still unknown or at least unrecognized. In spite of the finding by Avery *et al.* (1944) that DNA from one strain of *Pneumococcus* could transform the characteristics of another, the implications of this for genetic theory did not become immediately apparent. Therefore, at the close of the first half of the twentieth century, the view prevailed that genetic specificity probably resided in the protein part of the nucleoprotein complex (J. H. Taylor, 1965).

## III. DNA and the Molecular Gene

Notwithstanding the general lack of appreciation regarding the significance of DNA, a considerable amount of pertinent information had become available. Thus, Boivin *et al.* (1948) demonstrated that the amount of DNA in the somatic cells of various animal species was constant and twice as high as the amount of DNA in the sperm. Essentially similar results for plant nuclei were obtained by Swift (1950), although multiple values due to polyploidy were more frequently found in this material.

A major obstacle standing in the way of DNA being readily accepted as the genetic material was a lack of precise knowledge of the arrangement of its component parts. It had long been known that deoxyribonucleic acid contains a pentose sugar, deoxyribose, phosphoric acid, as well as two purines, adenine and guanine, and two pyrimidines, thymine and cytosine. Four different nucleotides are thus possible, each containing one purine or pyrimidine. Chargaff (1950) showed that the proportion of purine and pyrimidine bases varied in the DNA from different species; however, in each species, the ratios of adenine to thymine and of guanine to cytosine were nearly equal. It was on the basis of this finding, combined with X-ray refraction data by Wilkins *et al.* (1953), that Watson and Crick (1953) arrived at the model of the double helix, in which two polynucleotide chains, each consisting of a sugar-phosphate backbone, are held together by hydrogen bonds between purine and pyrimidine bases, either adenine paired with thymine or guanine paired with cytosine. Thus the two chains are complementary, and DNA duplication merely requires them to separate, with each acting as a template for a new chain.

The sequence of nucleotides along the DNA molecule determines the genetic specificity, three bases being required to code for one amino acid (see Watson, 1970). During this process, the information on the DNA is transcribed onto RNA nucleotides, which then act as templates for the assembly of amino acid sequences into proteins. Since most polypeptides contain between 200 and 600 amino acids, most genes contain between 600 and 1800 nucleotides. Mutations are changes in the sequence of base pairs.

Human adult hemoglobin, hemoglobin A, contains about 600 amino acid residues, but these are coded for by two genes, which are responsible for the formation of two $\alpha$ and two $\beta$ chains, respectively. Sickle cell disease is due to an abnormal hemoglobin, hemoglobin S, which is inherited as a Mendelian recessive, while the heterozygous carrier is said to have the sickle cell trait (Neel, 1949). The difference between normal hemoglobin and hemoglobin S is due to a single amino acid substitution in the $\beta$ chains, where one glutamic acid molecule is replaced by valine (Ingram, 1957). Furthermore, the change from glutamic acid to valine is explicable on the basis of a single base substitution in one nucleotide (Crick, 1966).

Many other abnormal human hemoglobins are known, which can be explained on a similar basis (Weatherall and Clegg, 1969). All these structural hemoglobin variants are examples of point mutations in the strict sense, that is substitutions in a purine or pyrimidine base giving rise to a changed amino acid. The term "mutation," however, is usually used in a much wider sense. In addition to base substitutions, deletions as well as insertions of single bases are included under the heading of a point mutations, and even chromosomal aberrations and numerical changes are considered as "mutations" (Röhrborn, 1970). In a large proportion of mutations, no evidence as to the extent of the underlying genetic change is available. Vogel (1970) wrote as follows: " . . . in human genetics 'point mutation' is used to describe all genetic events which do not manifest themselves as microscopically visible chromosome changes. Such loose usage unfortunately makes all general deductions, such as the total number of mutations occurring within a generation, equivocal." In other species of animals and of plants, the term "mutation" is subject to a similar lack of precision.

The implications of this are particularly far-reaching, as it is becoming clear that not all of the DNA of higher organisms is coding for specific proteins.

In addition to the structural hemoglobin variants, other abnormal hemoglobins are known, which have a different genetic cause. The thalassemias are a heterogeneous group of disorders associated with a reduced rate of globin synthesis. Although it is clear that no single abnormal control mechanism can underlie the various types of thalassemias, the precise causative mechanism has not yet been elucidated. It is evident, however, that the correct rate of hemoglobin synthesis will be dependent on a strict control of the kinetics of erythropoiesis throughout the life of the organism, so that any cause which might result in an abnormal rate of red cell production might also be expected to lead to an abnormal synthesis of hemoglobin. Evidence will be presented in the following chapter (as well as in Chapter 7) that structural changes in chromosomes, of an order of magnitude which makes them visible under the microscope, or which are just below the level of visibility, may affect the rate at which cells divide.

We may assume that, in addition to the specific nucleotide sequences which code for specific proteins, the chromosomes of higher organisms contain stretches of DNA which are concerned with the control of the rate of cell proliferation.

# Chapter 4

## Heterochromatin

### I. Introduction

For many years, geneticists have divided the chromosomal material into two parts, euchromatin and heterochromatin, and there have been a number of recent discoveries which have validated this distinction. It is generally assumed that genes, at least actively functioning ones, are carried in the euchromatic parts of chromosomes, while the heterochromatic parts are devoid of active genes. Now, however, it seems likely that heterochromatin may turn out to be of special relevance in the architecture of the chromosomes of higher organisms. Although probably all such chromosomes contain heterochromatic regions, sex chromosomes tend to exhibit the properties of heterochromatin to a particularly striking degree. Many appear entirely heterochromatic, at least during certain stages of mitosis or meiosis, or contain large conspicuous heterochromatic parts. Even now the concept of heterochromatin is not yet fully understood, but even a partial knowledge is likely to help in formulating the right questions with regard to the genetics of sex differentiation.

Originally, the concept of heterochromatin arose out of its cytological characteristics, and these will now be briefly discussed.

## II. Heteropycnosis

Heterochromatin was first described in liverworts and mosses, which are haploid during the major part of their life cycle. The liverwort, *Pellia epiphylla*, is hermaphrodite and has nine chromosomes in its haploid cells. Heitz (1928) found that during early prophase, the major part of one of these chromosomes, as well as shorter segments of four other chromosomes, were more condensed and, therefore, more intensely stained than the other chromosomal regions. The intensely staining regions could not be distinguished during metaphase, but during telophase they retained their chromosomal characteristics, while the rest of the chromosomal regions lost their cytological identities. The darkly staining chromosome parts persisted throughout interphase. In another species, *Pellia neesiana*, the sexes are separate, and female plants contain a large X chromosome, which is absent in male plants. The X chromosome appeared intensely stained in prophase and interphase cells.

The differential staining capacity of the X chromosome of insects, when compared with the rest of the chromosomes during certain stages of spermatogenesis, had already been described by a number of authors (see Chapter 1); Gutherz (1907) called the phenomenon of differential condensation and staining capacity, which in most cases implies a more highly condensed state and deeper staining ability, "heteropycnosis."

Heitz (1928) considered that the tendency toward heteropycnosis was inherent in the chromosomal regions themselves. He coined the term "heterochromatin" to denote the substance of those chromosomal regions which have a tendency to become heteropycnotic; the substance of the rest of the chromosomal regions, which became unrecognizable in interphase, he called "euchromatin." Although it is still not possible to define either of these substances in a precise way, recent results obtained with autoradiography and other new techniques have corroborated the original distinction between heterochromatin and euchromatin to a remarkable degree (this chapter, Sections XI–XIII).

Following his work on mosses and liverworts, Heitz (1933, 1934,

1935) studied the somatic chromosomes of *Drosophila melanogaster*. He concluded that in this species, the whole of the Y chromosome and about one-half of the X chromosome including the centromere were heterochromatic; in chromosomes 2 and 3, smaller pieces on either side of the centromere were found to be heterochromatic. In metaphase chromosomes, the heterochromatic regions appeared as constrictions, i.e., staining less intensely than the rest of the chromosomes. However, in prophase, the heterochromatic regions were more compact and more densely stained than the euchromatic parts. During interphase, the joint heterochromatic regions could often be seen as a darkly staining chromocenter.

The cytological features described thus far already suggest that, in different stages of the mitotic cycle, heterochromatin is out of phase with respect to euchromatin. This is best seen in interphase and in prophase, when heterochromatic regions are more highly condensed and more intensely stained than euchromatic ones. During mitotic metaphase, on the other hand, all chromosomes are in a state of maximum condensation. Certain heterochromatic regions may appear as lightly staining gaps at this stage; in mammalian chromosomes, heterochromatic blocks are characterized by close apposition of the chromatids in metaphase (Schmid, 1967).

## III. Genetic Effects

Before the discovery of the chromosomal nature of the salivary gland chromosomes in the larvae of *Drosophila melanogaster,* so-called cytological linkage maps were constructed correlating specific translocations and duplications, as seen in metaphase chromosomes, with the phenotype of the flies caused by the presence of marker genes (Dobzhansky, 1932). When the results were compared with the usual type of genetic linkage maps obtained by counting the proportion of cross-overs between linked genes, it became apparent that the order of genes obtained by the two methods was the same, but that the spacing of the genes along the chromosomes appeared to differ. The most striking finding was that in about one-half of the X chromosome, as seen in metaphase preparations, at the centromeric end, only a single gene, *bobbed* (see this chapter, Section V), could be located. In other words one-half of the X chromosome seemed to be genetically "inert" (Dobzhansky, 1932), and it was this part of the X chromosome, which Heitz (1933) found to be heteropycnotic. Chromosomes 2 and 3 were found to contain shorter inert regions, and although their exact location could not be determined at the time, Heitz postulated that there might be a general correspondence between cytological heteropycnosis and genetic inertness.

Regarding the absence of Mendelian genes, the hypothesis has withstood the test of time. Nevertheless, it has become clear that heterochromatic chromosomes did not leave the phenotype entirely unaffected. The Y chromosome in *Drosophila melanogaster* is of particular interest in this connection. Although in *Drosophila* the Y chromosome is not required for male determination (Chapter 1), males which do not possess a Y chromosome are sterile, having immotile spermatozoa. Stern (1929) located several regions on both arms of the Y chromosome, all of which are required to ensure motility of the sperm. These regions have been known as "fertility factors." A detailed analysis of the Y chromosome in *Drosophila* by Hess and Meyer (1968) has revealed that this chromosome forms lampbrushlike loops by an unfolding of its DNA in the spermatocytes of larvae. The loops are active in RNA synthesis, and if a segment of the Y chromosome containing a loop-forming site is missing, males carrying this chromosome are sterile. It seems likely now that the loop-forming sites are the fertility factors described earlier (Chapter 7, Section VIII).

Although lampbrushlike loops are confined to the nuclei of spermatocytes and their effect apparently confined to the sperm, the Y chromosome of *Drosophila* may affect other characters in more subtle ways. For instance, Mather (1944) showed that Y chromosomes of different strains were associated with differences in the mean numbers of sternopleural chaetae, or bristles. The experiments, moreover, suggested that parts of the Y chromosome responsible for this variation could form exchanges with the X chromosomes, thus implying that the heterochromatic regions of the X chromosome, too, may affect the number of bristles formed.

Heterochromatic regions are not the only ones which affect quantitatively varying characters, since Mendelian genes in addition to their more obvious qualitative effects, are known to result in quantitative variation (Chapter 2, Section II). Thus, whereas genes situated in euchromatin have specific qualitative as well as nonspecific quantitative effects, heterochromatin appears to have quantitatively varying effects only. On the other hand, changes in the heterochromatic regions tend to be far better tolerated than alterations in euchromatic chromosomes, which are likely to be much more detrimental to the organism. For this reason, heterochromatin may be particularly well fitted for the task of providing quantitative variation between related organisms, as shown by Brown (1966) for the tomato and the petunia, both of which are members of the same family of flowering plants, the Solanaceae. Schmid and Leppert (1968) found that different species belonging to the same subfamily of rodents, the Microtinae, differed in as much as 28% of their DNA contents per nucleus and this was thought to be due to differences in the amount of structural heterochromatin (this chapter, Section XIV). Indeed, differences in the arrangement of heterochromatic segments can occur within the same species. D. L. Southern (1970) described different arrangements of heterochromatic segments, as seen in meiotic prophase, in different populations of *Metrioptera brachyptera,* a species of long-horned grasshoppers, Tettigoniidae. The karyotype of the red fox (*Vulpes vulpes*) may contain between one to six tiny chromosomes, which are presumed to be heterochromatic (Buckton and Cunningham, 1971).

## IV. *Minutes* and Homoeotic Mutants in *Drosophila*

*Minutes* have been defined as a class of genes on different chromosomal loci of *Drosophila melanogaster,* which in the heterozygous state produce smaller (i.e., shorter and finer) bristles and lengthen the time of development of the flies; they are lethal in homozygous dose (Lindsley and Grell, 1968). Many of them are due to chromosome deletions. The effect they produce may be modified by additional Y chromosomal material. Goldschmidt (1955) was of the opinion that the *Minute* phenotypes are caused by quantitative changes in heterochromatin, though he admitted that such a claim is virtually impossible to substantiate. More recently, Farnsworth (1965) studied the growth of growth of larvae of ten different *Minutes* on chromosomes 2 and 3 by estimating total soluble protein; the cytochrome oxidase activity of mitochondria was also measured. In all *Minutes*, growth was greatly retarded throughout the larval period, but the time of maximum retardation varied in different *Minutes*. Farnsworth postulated that difficulties in oxidative phosphorylation might be the cause of the observed abnormalities and results obtained on isolated mitochondria supported this hypothesis (Farnsworth and Jozwiak, 1969). Abnormalities associated with cytochrome b-c sites of phosphorylation were found in all five *Minute* mutants which were assayed for this purpose. Thus, the impairment of protein synthesis and of growth might be due to a low level of ATP available in the cells of the *Minutes*. Fahmy and Fahmy (1966) have shown that the effects of DNA and other macromolecules, when injected into adult males or fed to larvae, were mainly in the production of *Minutes* and other small chromosomal deletions and that most of these were in heterochromatic regions.

The so-called homoeotic mutants in *Drosophila* are another example illustrating the possible importance of changes in heterochromatin on processes of development and differentiation. Homoeotic changes have been defined as "substitution of serial (segmental) homologs for each other or for part thereof" (Goldschmidt *et al.*, 1951). The subject will be discussed in somewhat more detail in Chapter 7 (Section II). In the present context, the mutant *podoptera*, in which the wings are changed into leglike structures, is of particular relevance. Although the condition is genetically determined, it was found that it was not caused by a particular gene locus, and the abnormal phenotype did not segregate in a clear-cut Mendelian manner. On the other hand, various heterochromatic regions of chromosomes were found to alter the expression of *podoptera*. The Y chromosome was particularly effective; in some strains, females would show the *podoptera* phenotype only if they carried a Y chromosome.

The experimental production of homoeotic changes has been described by Hadorn and his associates (Hadorn, 1968; Wildermuth, 1970), who called the process "transdetermination." This topic, also, will be discussed in Chapter 7 (Section V). Of special interest is the finding that transdetermination occurs only following high mitotic rates. It is thus possible that the so-called homoeotic mutants may also be dependent on the rate of cell division, and that this may be affected by the presence of a particular heterochromatic region.

These findings may be of significance in relation to the process of sex differentiation in mammals, which depends on whether a potentially hermaphrodite gonad develops into a testis or an ovary (Chapter 5). In mammals it has been postulated that the ovary represents the basic gonadal development and that testicular differentiation depends on additional mitotic divisions, normally mediated by the presence of a Y chromosome. There is evidence, which admittedly is tentative at present, that in both mammals and *Drosophila*, organ substitution during embryogenesis may be brought about by changes in chromosomes or chromosomal regions, which show heterochromatic properties, while the findings on experimental transdetermination indicate a connection between organ substitution and mitotic rates. These different sets of data strengthen the hypothesis that differences in heterochromatin may result in different rates of growth and that these, in turn, may give rise not only to quantitative variation but in some cases also to qualitatively different end products (Chapter 2, Section III).

## V. Nucleolar Organizers

Nucleoli are organelles present in interphase nuclei. They are built up by specific chromosomal regions, which appear to be heterochromatic. Heitz (1931, 1932) already commented on the relationship between nucleoli and satellite-bearing chromosomes in many species of plants. In maize (*Zea mays*), there is a single satellite-bearing pair of chromosomes, which is strikingly associated with the nucleolus (McClintock, 1933), while in man there are five pairs of satellited chromosomes and these are thought to be connected with the formation of nucleoli (Ohno *et al.*, 1961). The cytological appearance of satellites in metaphase—faintly staining, puffy structures separated from one end of the chromosome by means of a secondary constriction—suggests their heterochromatic nature. In *Drosophila melanogaster,* the heterochromatic ends of the X as well as of the Y chromosome act as nucleolar organizers (Kaufmann, 1934; Dobzhansky, 1944; Cooper, 1959). This assumption, which was based on cytological evidence for many years, was confirmed by Ritossa and Spiegelman (1965), using the DNA/RNA hybridization technique. The amount of DNA which is complementary to ribosomal RNA was found to be equal in normal male and female flies, thus suggesting that the nucleolar organizers of the X and the Y chromosome make equivalent contributions of this type of DNA. Subsequently, Ritossa *et al.* (1966) found that four different *bobbed* "mutants" were all partially deficient in this DNA. The authors suggest that the *bobbed* locus is, in fact, the nucleolar organizer and that the genetic basis of *bobbed* is a partial deletion of the DNA in the nucleolar organizer. Phenotypically, *bobbed* has long been recognized by its effect of shortening the bristles in homozygous females (see Lindsley and Grell, 1968). Hemizygous males have normal bristles as do XXY females, whereas XO males show the effect in an extreme form, and so it had therefore been assumed that the Y chromosome carries the normal allele of *bobbed.* It now appears that *bobbed,* which has long been regarded as a major gene situated in heterochromatin is, in fact, a deletion, which is too small to be seen cytologically; Therefore, the size and shape of bristles has been shown to be determined by a deletion of a heterochromatic region, which in turn results in a deficiency of a particular type of RNA. This finding prompts the question whether perhaps other genes affecting bristles, and possibly the shapes and sizes of other organs, may in reality be small chromosomal aberrations.

The African water frog, or clawed toad, *Xenopus laevis,* is a species of vertebrates which has provided further information on the relationship of nucleoli and the synthesis of ribosomes. This species normally has two nucleoli, which are visible by phase contrast microscopy in interphase nuclei. However, a female described by Elsdale *et al.* (1958) had about 5% of her offspring with only one nucleolus per cell and, when two of these were mated, their offspring was composed of individuals with two, one, and no nucleoli in a 1:2:1 ratio. While frogs with a single nucleolus appeared to be normal, those having none died as young larvae. Staining with pyronin led to the impression that the RNA contents in the cells of these larvae was reduced, while cytological evidence suggested that animals with one nucleolus (1-nucleolated) lacked one secondary constriction or nucleolar organizer, and larvae with no nucleoli (0-nucleolated) lacked two, in the majority of diploid cells (Kahn, 1962). The absence of ribosomal RNA synthesis in 0-nucleolate larvae was established by sucrose density gradient centrifugation, which showed that both 28 S and 18 S ribosomal RNA failed to be formed (Brown and Gurdon, 1964). These variants show typical Mendelian ratios in the progeny between crosses, and they have, therefore, been called "mutants," although undoubtedly they represent different deletions of the nucleolar organizer. It will be recalled that in *Drosophila*, these deletions result in the *bobbed* phenotype, but *Xenopus* has been less well studied for minor variants of morphology and none have been described in this case.

## VI. Repetitive DNA Sequences

We have seen that the nucleolar organizer is a stretch of chromosome which is responsible for the formation of the nucleolus, which in turn, produces ribosomal RNA. Ribosomes, which contain proteins as well as RNA, play a fundamental, role in protein synthesis by attaching to messenger RNA (de Man and Noorduyn, 1969).

During the last years it has become apparent that a considerable proportion of the DNA of higher organisms consists not of unique nucleotide sequences, but of sequences which are highly repetitive, often many thousands of times. One of the properties of this type of DNA is that when DNA is carefully separated into single strands, the two strands from highly repetitive DNA reassociate far faster than the DNA compound of unique sequences (Britten and Kohne, 1969a,b). A large proportion of the most highly repetitive sequence of mouse DNA can be separated from the remainder of the DNA by centrifugation in a cesium chloride density gradient, when it forms a minor component, or satellite band (Flamm *et al.*, 1967; Walker *et al.*, 1969). The composition of satellite DNA was found

to differ in closely related species of rodents. The different sequences must, therefore, have originated since the groups
separated (E. M. Southern, 1970).

The nucleolar organizer consists of highly repetitive DNA sequences for which a genetic function is known. The significance
of all the other such sequences is still being debated.

## VII. B Chromosomes

In some species of animals and plants, a proportion of individuals
contain one or more supernumerary, or accessory, chromosomes,
which are not part of the normal karyotype. They are generally
referred to as B chromosomes in contrast to the chromosomes of
the standard set, which are called A chromosomes. Müntzing (1967)
lists the properties of B chromosomes as follows:

1. They occur only in a proportion of individuals in a species
and are, therefore, not necessary for viability. In different populations of rye (*Secale cereale*), from 0 to 90% of individual plants
were found to carry B chromosomes.

2. The genetic effects of B chromosomes tend to be slight. In
rye and some other species, B chromosomes are associated with
reduced fertility and often with reduced vigor.

3. In various species, there are special mechanisms to maintain
or increase the number of B chromosomes, such as directed nondisjunction in meiosis preceding the formation of either male or female
gametes.

4. B chromosomes are usually smaller than the rest of the chromosomes and do not show any homology with them.

In addition, the presence of B chromosomes has been shown
to affect the number and particularly the variance of chiasmata
present in the A chromosomes in meiosis. Although, in general,
the effect of B chromosomes on this and other characteristics is
roughly proportional to their number, an odd number of B chromosomes was found to have a greater effect than when they were
presented in even numbers (Jones and Rees, 1969).

The relative genetic inertness of B chromosomes led to the sus-

which the female is the heterogametic sex. The same applies to the silkworm (*Bombyx mori*), in which a similar sex difference was discovered a few years later (Frizzi, 1948). The darkly staining body present in females was interpreted as the heteropycnotic Y(W) chromosome.

Soon afterward, Barr and Bertram (1949) published their discovery of a darkly staining body present in the nuclei of female cats but absent in nuclei from males. The body was first seen in the nuclei of neurons, where it was associated with the nucleolus. Subsequently, it was described in the nuclei of many tissues and in many species of mammals including man. In most of these nuclei, the body appeared to lie against the inner side of the nuclear membrane and was not associated with the nucleolus. The original term "nucleolar satellite" became, therefore, inappropriate and the body became known as "sex chromatin" or Barr body. The simplicity of the sex chromatin test has had a profound affect on the development of human cytogenetics in general. It has also focused attention on the dissimilar behavior of the two X chromosomes in somatic cells of mammalian females.

Much more recently, it has become possible to distinguish human Y chromosomes in interphase cells by means of a fluorescent staining technique. This will be discussed later in this chapter (Section XII).

A typical Barr body is highly condensed, with a diameter of about 1 $\mu$m. It stains with Feulgen and other nuclear dyes and can also be seen by phase contrast microscopy in living cells. The majority of Barr bodies are seen adjacent to the nuclear membrane at the periphery of the nucleus; fewer Barr bodies appear in other situations within the nucleus and many of these are seen adjacent to a nucleolus. In proliferating cells, Barr bodies can be seen before, during, and after DNA synthesis (Fig. 4.1).

That a Barr body originates from one of the two X chromosomes present in females was first suggested by Ohno *et al.* (1959). The assumption is supported, first, by the numerical relationship between the number of X chromosomes in the karyotype and the maximum number of Barr bodies in interphase nuclei; in diploid cells, these follow the $n - 1$ rule. At least two X chromosomes must be present for one Barr body to be formed and if more

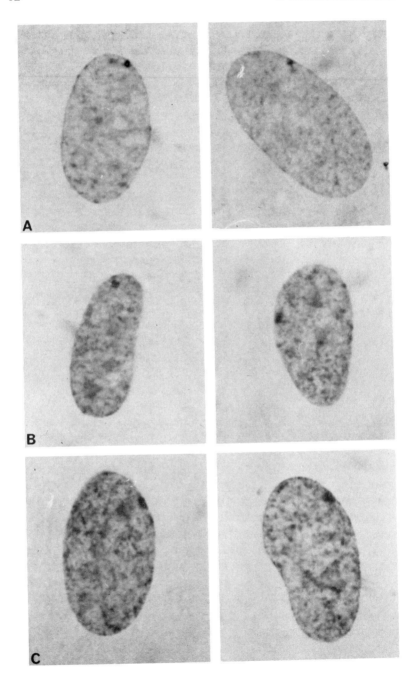

than two X chromosomes are present, more than one Barr body may be formed. The presence of a Y chromosome does not seem to affect Barr body formation to a measurable degree, but polyploidy does. The numerical relationship between numbers of Barr bodies, X chromosomes, and sets of autosomes is given by the following formula:

$$B = X - P/2$$

where $B$ is the maximum number of Barr bodies per cell, $X$ is the number of X chromosomes, and $P$ is the two sets of autosomes (Harnden, 1961). In triploid cells, the maximum number of Barr bodies is the same as in diploid cells with the same number of X chromosomes, but the proportion of cells showing this number of Barr bodies is less than in corresponding diploid one (Edwards et al., 1967).

The incidence of Barr bodies varies in different tissues, but the causes of this variation have been only partly established. The incidence of Barr bodies increases as the cells become more crowded (Klinger et al., 1968) and their nuclei become smaller (Mittwoch, 1967b; Issa et al., 1969). Older cells are more likely to contain Barr bodies than younger ones. In rapidly dividing tissues, the incidence of Barr bodies is lower than in cells in which mitoses have become rare (Therkelsen and Petersen, 1962). In early embryos, when cells are large and rapidly dividing, Barr bodies are not formed (Glenister, 1956; Austin and Amoroso, 1957; Park, 1957; Thorburn, 1964; Issa et al., 1969).

## X. Single Active X Chromosome Hypothesis

The fact that Barr bodies are not present in all cells in which they might theoretically be formed poses a problem in connection with the single active X hypothesis, as proposed by Lyon (1961,

*Fig. 4.1.* Barr bodies in nuclei of fibroblasts cultured from human female. Different stages of DNA synthesis; (A) $G_1$ phase, (B) S phase, and (C) $G_2$ phase. (Feulgen stain, photographed under standard conditions throughout, $\times$ 2200.) (From U. Mittwoch, and D. Wilkie, *Brit. J. Exp. Pathol.* **52**, 186, 1971).

1970, 1971, 1972) (usually referred to as the "Lyon hypothesis"). Its postulates may be subdivided into four parts:

1. The genes on the sex chromatin forming X chromosome are inactive.

2. Inactivation of the X chromosome occurs at random, so that paternal and maternal X chromosomes have an equal chance of being inactivated.

3. The process of inactivation occurs early in embryonic life.

4. Once inactivation has begun, the same X chromosome will always be inactivated in the descendants of each cell.

Evidence in favor of postulates (1) and (4) has come from the use of cultured fibroblasts (Davidson *et al.*, 1963; and see Krooth, 1969 for review.) Whereas cells from women who were heterozygous for genes on the X chromosome, e.g., two variants of glucose-6-phosphate dehydrogenase, were found to give rise to two types of clones, each one expressing only one allele, cells heterozygous for autosomal genes, e.g., lactate dehydrogenase, gave rise to only one type of clone expressing both alleles.

According to Deys *et al.* (1972), inactivation involves both arms of the X chromosome. Two out of 22 clones of fibroblasts derived from a woman who was heterozygous for a deficiency of phosphoglycerate kinase, which is known to be caused by a gene on the X chromosome, failed to show any phosphoglycerate kinase activity, while the other 20 clones had normal amounts. The authors favor the view that the gene for phosphoglycerate kinase is located on the long arm of the X chromosome, whereas the genes for glucose-6-phosphate dehydrogenase and hypoxanthine (guanine) phosphoribosyltransferase, which is also known to undergo inactivation, are thought to be situated on the short arm of the X chromosome.

Evidence suggesting that inactivation may occur even if no sex chromatin is formed has been given by Schwarzacher and Pera (1970). The short-tailed vole, *Microtus agrestis*, has giant sex chromosomes (Matthey, 1950). The X chromosome is about four times the length of the usual mammalian X chromosome and is thought to be a composite structure (Wolf *et al.*, 1965; Schmid *et al.*, 1965). About one-quarter of its length may be regarded as homologous with the standard mammalian X chromosome, and this part is euchromatic in the male, while in the female it is euchromatic in one X chromosome and heterochromatic in the other. The remaining three-quarters are heterochromatic in both sexes. The Y chromosome, which is almost as long as the X chromosome, is also heterochromatic. Interphase cells of both sexes may, therefore, contain large heteropycnotic bodies, though the extent to which they do so varies in different tissues. Schwarzacher and Pera (1970) found little uptake of tritiated uridine in the regions of the heteropycnotic bodies, as would be expected if these chromosomal regions failed to synthesize RNA. Similar regions of nonincorporation of tritiated uridine were said to be present in cells which did not show heteropycnotic bodies. It must, of course, be remembered that in cells of female *Microtus agrestis,* only about one-seventh of the heteropycnotic bodies corresponds to the usual type of sex chromatin body.

The results of cloning human fibroblasts, referred to above, as well as those from other species (see below), are evidence in favor of postulate (2) of the Lyon hypothesis that genes either on the paternal or on the maternal X chromosome may be inactivated, although they leave open the question whether the process is truly random. In some special cases it is likely that inactivation is not random. Thus, Polani *et al.* (1970) have presented evidence that in patients with Turner's syndrome who have a structurally abnormal X chromosome, this chromosome is always inactivated, while the normal X chromosome remains active. A number of such patients have become known with deletions either of the short arm or of the long arm of the X chromosome, who do not express the sex-linked red cell antigen $Xg^a$, although their fathers do. While the locus for $Xg^a$ may be either on the short arm or on the long arm of the X chromosome, it cannot be on both. Therefore, at least some of the $Xg(a-)$ patients must be carrying the antigen, which is assumed to be inactivated in a structurally abnormal X chromosome. Although random inactivation of the two X chromosomes followed by rigorous selection of only those cells in which the normal X chromosome is active remains a formal possibility, there is no evidence to support such a view. By contrast, the available evidence now suggests that the $Xg$ locus is not inactivated when present on a normal X chromosome. Red cells from women who are heterozygous for *Xg* do not fall into two separate classes with and without the antigen. The possibility that the *Xg*$^a$ antigen is not made by the precursor of the red cell but is merely absorbed from an unknown source was disproved when a pair of chimaeric twins was discovered, both of whom had two populations of red cells which segregated at the *ABO* as well as the *Xg* locus. The blood in each twin contained a mixture of $A_1B,Xg(a-)$ and $O,Xg(a+)$ cells (Race, 1973). This finding showed that red cells with and without the *Xg*$^a$ antigen can be produced in the same circulation. Furthermore, female patients with chronic myeloid leukemia have the usual female distribution of *Xg* blood groups (Lawler and Sanger, 1970). Provided the hypothesis of the clonal origin of chronic myeloid leukemia is correct, this finding suggests that in those cases in which the leukemia started from a cell with a late replicating X chromosome, the *Xg*$^a$ antigen is nevertheless expressed. In another type of tumor, leiomyoma of the uterus, women who were heterozygous for two variants of the sex-linked enzyme, glucose-6-phosphate dehydrogenase (*G-6-PD*) A and B, showed both variants in normal cells of the uterus but only one variant in the tumors (Linder and Gartler, 1965). This finding is evidence in favor of inactivation of the *G-6-PD* locus and also, to some extent, for the clonal origin of tumors; although, since tumors with the A as well as the B variant were found in all uteri, a number of distinct clones must have originated in each case.

Results of species crosses have provided strong evidence in favor of gene inactivation on the late replicating X chromosome. A mule is a hybrid between a female horse (*Equus caballus*) and a male donkey (*E. asinus*) and a hinny is the reciprocal cross (female donkey by male horse). The parental species are distinguished both by their diploid chromosome number—64 in the horse and 62 in the donkey—as well as by the shape of their X chromosome, that of the horse being more nearly metacentric. The two X chromosomes are cytologically distinguishable in the hybrid (Fig. 4.2). Although initially it was thought that either X chromosome had an equal chance of being late replicating (this chapter, Section XII) and thus presumably was genetically inactive (Mukherjee and Sinha, 1964), it subsequently became clear that the X chromo-

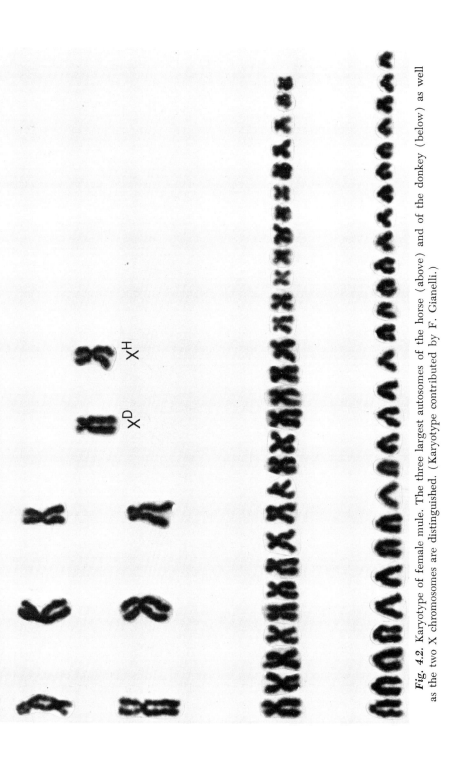

$X^D$ $X^H$

*Fig. 4.2.* Karyotype of female mule. The three largest autosomes of the horse (above) and of the donkey (below) as well as the two X chromosomes are distinguished. (Karyotype contributed by F. Gianelli.)

some of the donkey is late replicating in more than 50% of cases (Hamerton *et al.*, 1969a; Hamerton and Gianelli, 1970). These experiments were combined with the biochemical demonstration of the enzyme glucose-6-phosphate dehydrogenase, which is sex-linked in both the horse and the donkey. Moreover, the two species have characteristic electrophoretic variants, and, in the hybrid, by far the larger part of the enzyme showed the pattern characteristic of the horse. These results were confirmed by cloning (Hamerton *et al.*, 1971). Over 80% of cloned fibroblasts showed the *G-6-PD* variant of the horse. It is of interest to note that essentially similar results were obtained in both the mule and the hinny, which demonstrates that the apparently nonrandom inactivation in this species cross does not depend on a difference between paternal and maternal X chromosomes. It is not yet clear whether the results obtained reflect an originally nonrandom inactivation process, in which X chromosomes of the donkey were preferentially inactivated, or whether the excess of cells with active horse X chromosomes is a secondary result owing to better growth of these cells. The possibility of an organ-specific difference was suggested by Hook and Brustman (1971). The possibility must also be considered that different results may be obtained using, for instance, different breeds of horses.

Recent experiments on cloned cells from mules have provided additional support in favor of the hypothesis that genes on the late replicating X chromosomes are not expressed (Ratazzi and Cohen, 1972; Ray *et al.*, 1972). Autoradiography and *G-6-PD* typing were performed on the same clones and it was found that in those clones which expressed the *G-6-PD* variant of the horse, all the late replicating X chromosomes were those of the donkey, and vice versa. The data by Ratazzi and Cohen (1972) further showed that, in fibroblasts of the mule, the proportion of late replicating X chromosomes from the donkey, which was always greater than 50%, increased to nearly 100% with time in culture.

The preferential inactivation of paternal X chromosomes is suggested—although not yet conclusively proved—by the results of crossing different species and subspecies of marsupials (Sharman, 1971; Richardson *et al.*, 1970; Cooper *et al.*, 1971). Hybrids have been obtained between the euro (*Macropus robustus erubescens*) and the wallaroo (*M.r. robustus*), and between either of these subspecies as one parent and the red kangaroo (*Megaleia rufa*) as the other. Each of the three X chromosomes is cytologically distinguishable, in size as well as in shape, that of the wallaroo being the smallest while that of the red kangaroo the largest. In female hybrids, the paternal X chromosome was found to be late replicating in the large majority of cells, and this was associated with *G-6-PD* variants, which appeared to be exclusively of maternal origin.

The opposite situation has been found in species hybrids of gazelles (Wahrman, 1972). Hybrids between *Gazella gazella* and *G. dorcas* can be obtained only if *G. gazella* is the female and *G. dorcas* the male parent. The X chromosomes of the two species are distinguishable, since that of *G. gazella* is submetacentric while the X chromosome in *G. dorcas* is acrocentric. In two female hybrids, the submetacentric, i.e., maternal, X chromosomes were found to be late replicating in all analyzable cells.

These data suggest that when the X chromosomes are different the decision as to which is to be the late replicating one may not be random. On the other hand, the data on equine hybrids confirm the postulate that genes on the late replicating X chromosome are inactive.

Perhaps the most serious difficulty standing in the way of a full acceptance of the Lyon hypothesis is the pathological effect of abnormal numbers of X chromosomes in man (Chapter 5). Admittedly, the effects are less severe than if autosomes of similar size were involved, but different numbers of X chromosomes certainly have an effect on the phenotype. It is possible that just as B chromosomes may affect such basic parameters as mitotic cycle times without apparently contributing any specific gene loci, other chromosomes, also, may have an affect on the speed of cell proliferation, which is to some extent independent of the gene loci which may be present. Recent evidence obtained by Barlow (1972) supports this possibility. In fibroblast cultures, grown from females who were mosaics for two different X chromosome constitutions, cells containing a single X chromosome (45, X) were found to have shorter mitotic cycle times than cells with either two or three X chromosomes. Thus, a heterochromatic X chromosome may affect growth processes in the embryo, as well as in later stages, even though all, or most, of its genes are inactive. Quantitatively, this effect might be expected to be smaller in the case of a heterochromatic chromosome than in a euchromatic one, since it is well known that alterations in heterochromatic chromosomes are far better tolerated by organisms than any change in euchromatic chromosomes (Brown, 1966). It is likely, therefore, that if two or more X chromosomes are present, the genes of only one X chromosome are active, in accordance with the Lyon hypothesis, but that the presence of the additional X chromosomes has a quantitative effect on cell proliferation and related processes, thus producing a more or less marked effect on the phenotype.

## XI. Autoradiography

For many years, the idea of heterochromatin was opposed by geneticists who felt that the concept was too imprecise to merit serious consideration. That the topic occurs with increasing frequency in the modern literature is due to the development of autoradiography, as well as to more recent techniques of fluorescent and Giemsa staining, all of which gives results which confirm the concept of heterochromatin, as based on the cytological finding of heteropycnosis.

Autoradiography, which was developed by Doniach and Pelc (1950) is based on two sets of facts:

1. Synthesizing cells take up compounds containing radioactive isotopes of carbon, nitrogen, phosphorus, hydrogen, etc., indiscriminately with substances containing nonradioactive isotopes.

2. Once incorporated into the cells, the presence of the radioactive isotopes can be revealed by the application of radiosensitive films.

By means of autoradiography, Howard and Pelc (1953) showed that cells in the root tip of the bean, *Vicia faba*, take up sodium phosphate labeled with $^{32}P$ only during the S phase, prior to mitosis. Subsequently, Plaut and Mazia (1956) used thymidine, which is incorporated exclusively into DNA. This method was further refined by Taylor *et al.* (1957), who introduced thymidine labeled with tritium, $^3$H, which gives better cytological resolution than does $^{14}$C. Since then, autoradiography by means of tritiated thymidine has been widely used to investigate problems of chromosome replication in a large variety of organisms (Monesi, 1969).

It was by means of this technique that Lima-de-Faria (1959) demonstrated a difference in the time of uptake of tritiated thymidine between heterochromatic and euchromatic regions of chromosomes. This was found almost simultaneously in a grasshopper, *Melanoplus differentialis,* and in rye, *Secale,* a flowering plant. In *Melanoplus,* the single X chromosome of the male forms a heteropycnotic body during prophase of the first meiotic division, and this chromosome was found to incorporate tritiated thymidine later than the rest of the chromosomes. The chromosomes of rye have heterochromatic regions near the centromeres and Lima-de-Faria concluded that these regions also incorporated tritiated thymidine later than the euchromatic parts of the chromosomes. The late replications of heterochromatic chromosomes and the chromosomal regions have been found to be an almost universal attribute of these parts (Lima-de-Faria, 1969). In *Drosophila melanogaster,* the organism in which the classical cytogenetic studies on heterochromatin were originally performed, the followed chromosomal regions have been found to be late replicating: the entire Y chromosome, the centromeric portions of the X chromosomes and of autosomes 2 and 3, as well as the small autosome 4 (Barigozzi *et al.,* 1966). The results of autoradiography in this species thus show excellent agreement with earlier attempts at defining heterochromatin.

The first mammalian cells to be investigated by means of autoradiography were those of the Chinese hamster, *Cricetulus griseus.* Taylor (1960) found that in cultured cells from female hamsters, one of the X chromosomes replicates at the same time as the autosome, while the other X chromosome replicates later. The single X chromosome in cells from male animals replicates at the same time as the autosomes. Asynchronous replication of the two X chromosomes in female mammals has been studied in most detail in the human species (for instance, German, 1962, 1967; Schmid, 1963). In addition to normal males and females, large numbers

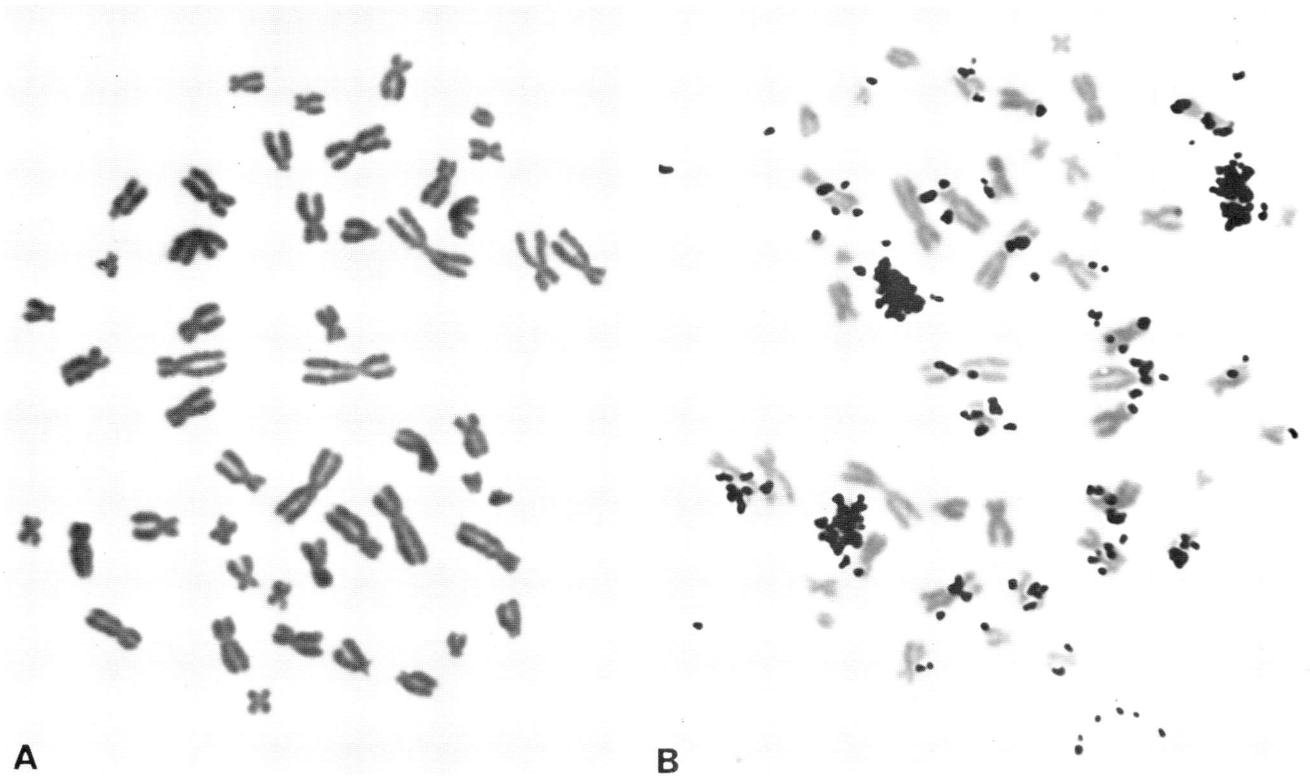

**Fig. 4.3.** Human chromosomes from patient with 49, XXXXY chromosomes labeled with tritiated thymidine late in S phase. (A) Before applying radiosensitive film. (B) After applying radiosensitive film. (From U. Mittwoch, *Sci. Amer.* **209,** 54, 1963).

of sex chromosome abnormalities are known in man, the most important being numerical and structural abnormalities of the X chromosome (Chapter 5). In patients with multiple X chromosomes, all but one are typically late labeling (Gianelli, 1963; Grumbach et al., 1963; Rowley et al., 1963). These findings lead to the almost inevitable conclusion that the late replicating X chromosomes form the sex chromatin bodies in interphase. The finding that structurally abnormal X chromosomes always replicate late (Polani et al., 1970) provides additional confirmation, since such X chromosomes are associated with sex chromatin bodies of abnormal size. In the case of the presumptive isochromosome of the large arm of the X chromosome (Xqi), Barr bodies and drumsticks are typically larger than normal; while in the case of deleted X chromosomes, Barr bodies and drumsticks are smaller than normal (see Mittwoch, 1967a).

The human Y chromosome, while not as obviously late replicating as the X chromosome, replicates later than autosomes 21 and 22, which are similar in size to the Y chromosome (Schmid, 1963; German, 1967). The Y chromosome of the mouse, *Mus musculus*, was found to be extremely late labeling in spermatogonial cells *in vivo* and late labeling could also be observed in cultured kidney cells. No clear-cut asynchrony of the sex chromosomes was evident in cells of the bone marrow, in which the $G_2$ period was on an average shorter than in the other cell types (Tiepolo et al., 1967). Basically similar results were found in the rat, *Rattus norvegicus*. The suggestion that differences in the pattern of DNA replication of individual chromosomes may be related to differences in mitotic cycle times opens up important possibilities regarding the role of chromosomes in controlling the rate of cell proliferation. This topic will be discussed further in Chapter 7. The Y chromosome is also late labeling in the rabbit, *Oryctolagus cuniculus* (Issa et al., 1969), and in somatic cells of the golden hamster, *Mesocricetus auratus* (Utakoji and Hsu, 1965).

The development of asynchrony of the X chromosome was investigated by Issa et al. (1969) in early embryos of the rabbit. The results suggest that late replication of one of the X chromosomes precedes Barr body formation. Barr bodies could not be detected until the embryos were aged 96 hours, when their cell number was estimated as between 160 and 380. A pronounced increase in the incidence of Barr bodies was observed over the next 24 hours. Late labeling of one of the X chromosomes could be detected in some of the metaphases of rabbit embryos aged 2–3 days (cell number 16–128). Kinsey (1967) did not find a late replicating X chromosome at this stage, although these were present in the older embryos. Hill and Yunis (1961) obtained even earlier embryos of the golden hamster, *Mesocricetus auratus,* and found that at the 8-cell stage, the two X chromosomes took up the label at the same time.

Autoradiography with tritiated uridine can be employed for studying RNA synthesis, while tritiated amino acids are used as precursors of protein synthesis. Results obtained by means of these techniques have given some indication that these synthetic processes tend to be inhibited when the DNA is in a condensed state (Monesi, 1965). The condensed X chromosome in interphase of mammalian females does not seem to synthesize RNA. By contrast it now seems likely that this chromosome synthesizes DNA while in the condensed state (see Schwarzacher and Pera, 1970).

## XII. Fluorescence Microscopy

The possibility of scoring Barr bodies in conjunction with the techniques of autoradiography have been a strong influence in the development of modern cytogenetics by greatly enlarging our knowledge of the behavior of the mammalian X chromosome, in both normal and abnormal situations. Until recently, however, there has been no simple method of estimating the number of Y chromosomes in the karyotype. The development of the fluorescent staining technique now makes it possible to do this in human cells. Moreover, results obtained in this way are likely to shed further light on the properties of heterochromatin (Pearson, *et al.* 1971).

The technique of utilizing fluorescent alkylating agents for the purpose of differential staining of specific chromosomal regions was developed by Caspersson *et al.* (1969a,b). By treating cells with quinacrine or quinacrine mustard and observing them in an ultraviolet high resolution fluorimeter, the authors found that the dye binds to specific regions in the chromosomes of *Vicia* and *Trillium*, producing brightly fluorescing bands. The position of the fluorescence corresponded to heterochromatic regions as visualized by cold treatment. Subsequently, Zech (1969) applied the technique to human chromosomes and found that the Y chromosome stood out in metaphase by the brightly fluorescing ends of its long arm (Fig. 4.4). Soon afterward, Pearson *et al.* (1970) showed that the human Y chromosome could be identified as a brightly fluorescing spot in interphase nuclei by staining with a fluorescent acridine derivative, quinascrine hydrochloride. Thus, the presence

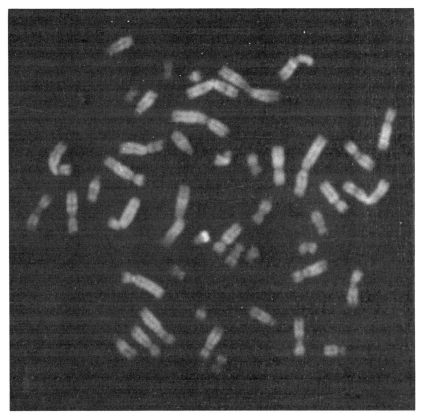

*Fig. 4.4.* Chromosomes from human male showing brightly fluorescing Y chromosome near center of cell. (Photograph contributed by P. L. Pearson.)

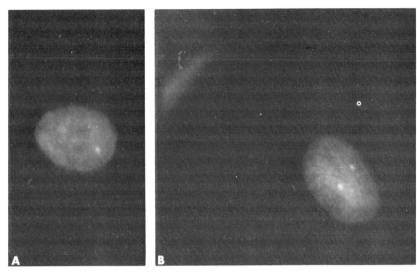

*Fig. 4.5.* Buccal mucosa cells from human males, fluorescent staining. (A) From normal male, showing a single Y body; (B) from YY male, showing two Y bodies. (Photographs contributed by P. L. Pearson.)

of one, or more than one, Y chromosome can now be ascertained in such easily obtainable cells as buccal mucosa and lymphocytes (Fig. 4.5). The fluorescent Y body can also be seen in sperm (Sumner *et al.*, 1971a) and thus provides the first successful means of separating the X- and the Y-bearing sperm, although in a fixed condition. Its identification in cells of amniotic fluid is likely to improve the accuracy of prenatal sexing techniques (Rook *et al.*, 1971).

The autosomes also have characteristic fluorescing patterns by which they can be distinguished from each other, although none of the autosomal fluorescent bands are as large as that seen in the Y chromosome. The heteropycnotic X chromosome does not fluoresce (George, 1970; Manolov *et al.*, 1971). However, Mukherjee *et al.* (1972) reported the fluorescence of both X and Y chromatin. The X and Y bodies were said to be distinguishable by size, the Y being smaller than the X body.

In general, chromosomal regions which replicate late also fluoresce brightly with quinacrine dyes, although a few exceptions have also been noted. Vosa (1970) observed that in *Drosophila melanogaster,* the centromeric regions of the X chromosome and of chromosomes 3 and 4 were both fluorescing and late replicating, whereas the centromeric region of chromosome 2 was late replicating but did not fluoresce. The Y chromosome, which appears to be entirely heteropycnotic and late replicating, showed several nonfluorescing bands. In human chromosomes, the intensity of labeling with tritiated thymidine and the distribution of fluorescing regions was compared by Ganner and Evans (1971), with particular reference to chromosomes 1, 13, 14, 15, 17, 18, 19, 20, 21, and 22. They found that, in general, chromosomal regions which are late replicating also show bright fluorescence. However, the late replicating X chromosome in cells of female origin is late replicating but does not fluoresce, while the centromeric regions of chromosomes 1, 9, 16, and 22, which are late replicating, showed a dull fluorescence. It seems that brightly fluorescing regions are also late replicating, but not all late replicating chromosomal regions show bright fluorescence. A precise correlation between late replicating chromosomal regions and intense quinacrine fluorescence was noted by Ellinson and Barr (1972) for the chromosomes of *Samoaia leonensis,* a member of the family Drosophilaedae whose chromosomes are twice the size of that of *Drosophila melanogaster.*

The possible causes underlying the differential fluorescent patterns are still being debated. A relationship betwen intensity of fluorescence and guanine contents was suggested by Caspersson *et al.* (1969b) and by Rowley and Bodmer (1971). Comings (1971) has put forward the view that the fluorochrome may be sensitive to nonhistone proteins, which are bound to certain parts of the DNA. Ellinson and Barr (1972) suggest that intense fluorescence may be due to regions which are especially rich in adenine and thymine.

## XIII. Selective Giemsa Staining

Yet another recently developed technique for the differential staining of heterochromatin involves a partial denaturing of the

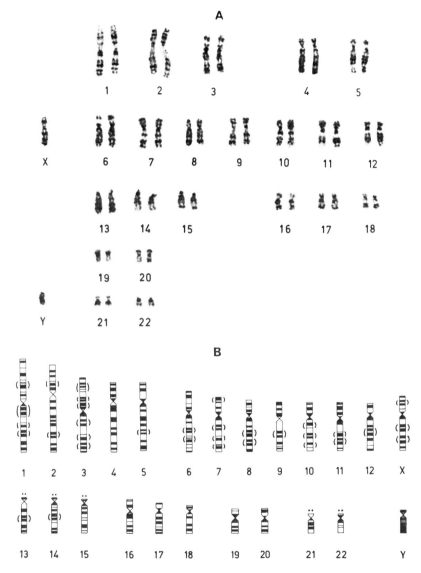

*Fig. 4.6.* Chromosomes from human male, showing banding of heterochromatic regions following selective Giemsa staining. (A) Photograph; (B) schematic diagram. (From Schnedl, 1971.)

chromosomes and subsequent staining with Giemsa. Used in the conventional way, Giemsa would stain the entire chromosome, but following the denaturing of DNA, the heterochromatic regions containing repetitive DNA will be selectively stained, since these reassociate faster than the unique DNA. By denaturing the DNA of mouse chromosomes with a solution of sodium hydroxide and hybridizing with radioactive DNA and subsequent staining with Giemsa, Pardue and Gall (1970) were able to show cytologically that mouse satellite DNA is situated in the centromeric regions. Denaturing with sodium hydroxide prior to Giemsa staining was also used by Arrighi and Hsu (1971) to stain the centromeric regions of human chromosomes and the method was subsequently modified to show up the characteristic banding patterns of individual chromosomes (Schnedl, 1971; Drets and Shaw, 1971; Fig. 4.6). Sumner et al. (1971b) introduced the use of saline as a denaturing agent, while Seabright (1972) used proteolytic enzymes. The results obtained depend to some extent on the technique used. In chromosomes of the rat (*Rattus norvegicus*), Schnedl and Schnedl (1972) obtained Giemsa banding patterns after denaturing briefly with 0.002 $M$ sodium hydroxide, whereas with a slightly longer treatment with sodium hydroxide, or a more concentrated solution (0.07 $M$), only the centromeric regions were darkly stained. The amount of darkly staining centromeric material also differs in different chromosomes; in human chromosomes, chromosomes 1, 9, and 16 have particularly well-marked centromeres (Evans et al., 1971). A certain amount of variation can also be observed between members of a homologous pair of chromosomes. The long arm of the human Y chromosome stains darkly with Giemsa. As in the case of fluorescent staining, the significance of the Giemsa bands is at present a matter of conjecture (Evans and Sumner, 1973).

The darkly staining regions make up roughly one-half of the chromosomes, which suggests that the proportion of heterochromatin in the karyotype is likely to be larger than had originally been assumed. Also, a recent report by Saunders et al. (1972) suggests that in human chromosomes, satellite DNA may be present in the apparently euchromatic regions as well as around the centromeres.

## XIV. Constitutive and Facultative Heterochromatin

The contrasting behavior of the two X chromosomes in female mammals has drawn attention to the fact that the characteristic features of heterochromatin are not necessarily the result only of the organization of the chromosome itself, but that they may be imposed upon it by outside factors such as the presence of a homologous chromosome in the karyotype. This state of affairs, in which one partner becomes heterochromatic while the other remains euchromatic has been called "facultative heterochromatinization," and the heterochromatic chromosome is said to contain facultative heterochromatin. By contrast, heterochromatic regions which appear to be equal in two homologous chromosomes—as, for instance, the regions adjacent to the centromeres—are said to be composed of "constitutive (or structural) heterochromatin" (Brown, 1966; Schmid, 1967). Whereas in facultative heterochromatin, potentially functional gene loci are known to be present, there is no evidence for the existence of such loci in constitutive heterochromatin; they may have disappeared during the process of evolution.

Facultative heterochromatization is not confined to the X chromosome of mammals. In male mealy bugs, *Gossyparia spuria*, the entire paternal set of chromosomes becomes heterochromatic and these chromosomes are eliminated from the cells which give rise to the sperm; thus, a male mealy bug transmits only the maternal set of chromosomes (Schrader, 1929; Hughes-Schrader, 1948; Brown, 1966). It is of interest that in very young embryos there is no sign of the heterochromatic nature of the paternal chromosomes. This case, therefore, provides a clear parallelism with the development of asynchrony of one of the X chromosomes in mammalian females. Moreover, it has been reported that heterochromatic blocks are not visible in very early embryos of *Drosophila melanogaster* (Cooper, 1959), and this suggests that at least some constitutive heterochromatin does not develop heterochromatic properties until some time during embryogenesis. It has been suggested that heterochromatin may be better regarded as a state that a substance (Brown, 1966), and White (1954) believes that heterochromatin and euchromatin may be two opposite states in a continuous series, between which intermediate states may occur. Facultative heterochromatin may be an obvious candidate to take up such an intermediate position.

It is possible that heterochromatinization is dependent on a prolonged mitotic cycle time, although at present this is little more than guesswork. Clearly, an understanding of the processes which bring about heterochromatinization in the embryo will be necessary not only in order to solve the problem of what heterochromatin is and does, but also in order to settle the wider issue of the role of the chromosomes in embryogenesis.

In spite of the formidable tasks ahead, the steadily increasing weight of evidence has removed any lingering doubt as to the biological reality of heterochromatin, both facultative and constitutive.

# Chapter 5

## The Nature of Sex Differentiation
## with Special Reference to Vertebrates

### I. Introduction

In vertebrate development, differences between males and females are first seen at the level of the gonads, and this is often referred to as the primary sex difference. Secondary sex differences include those of the genital ducts and the external genitalia, the mammary glands of mammals, as well as elements in the pituitary controlling various parts of the reproductive cycle. The development of secondary sexual characters is normally determined by the nature of the gonad, and, for this reason, it is the process of gonadal differentiation with which we are primarily concerned.

Differences in the morphology and physiology of males and females are best known in man, although many mammalian species do not lag far behind in this respect. Striking sex differences may also be seen in many nonmammalian vertebrates, but in some of these the separation between the sexes is not as rigid as in mammals.

Individuals having either male or female reproductive organs, but not both, are said to be "unisexual," "dioecious," or "gonochoristic." If they have both male and female reproductive organs, they are said to be "bisexual," "monoecious," or "hermaphrodite." The terms "monoecious" and "dioecious" are usually confined to plants, while "unisexual" and "bisexual" are confusing. Hermaphroditism may represent the normal state within a species, or it may refer to an abnormality of individuals in a species whose members are normally gonochoristic. The term "intersex" denotes various abnormalities of sexual development.

We shall begin by examining the sex difference in man and other mammals.

## II. Germ Cells

Males and females differ in a large number of, if not all, variables that are amenable to measurement. Such quantities as body height and weight, the growth of bones, and the distribution of fat, as well as numerous physiological variables, including blood pressure and basal metabolic rate (heat produced per unit body surface), all show different distributions in the two sexes (Tanner, 1962). However, the most far-reaching differences clearly occur in relation to reproduction.

The essential fact of sexual reproduction consists of the fusion of two germ cells, which though homologous, differ greatly in appearance (Fig. 5.1). The most striking difference is that of size, but the cells also vary greatly in structure. A human egg is a spherical cell with a diameter of 130–140 $\mu$m. Thus the volume of the egg is about 2,000,000 $\mu$m$^3$, which makes it one of the largest cells of the body, and when placed against a dark background it is just visible to the naked eye. Most cells have volumes between 200 and 15,000 $\mu$m$^3$ (Austin, 1961). However, a motor neuron in a large mammal would have a volume of about 10,000,000 $\mu$m$^3$. This is because the axon (cytoplasmic process) of such a cell may have to reach from the nervous system right down to the end of a limb.

The large size of egg cells is due mainly to deposits of yolk in the cytoplasm. There is little or no correlation between egg size in mammals and the size of the animals. Thus, the egg of the horse is about the same size as that of the rabbit and both are only slightly smaller than the egg of the whale.

In contrast to the egg, the sperm is the smallest cell of the

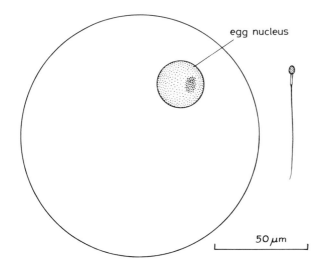

egg nucleus

50 μm

*Fig. 5.1.* Mammalian egg and spermatozoon, drawn to the same scale.

body; a human sperm cell has a volume of about 30 $\mu m^3$. This means that the volume of an egg cell is about 85,000 times that of a sperm. The difference in morphology is almost equally striking. The sperm is a highly differentiated cell, whose main components are a head and a tail. The head is mainly composed of the cell nucleus, which is very much condensed. The long tail consists of two main parts, the middle piece and the end piece. The detailed morphology of spermatozoa has been extensively studied by both light and electron microscopy (Bishop, 1961).

Notwithstanding the many differences shown by mature egg and sperm cells, both types of cells develop according to a common plan, as was demonstrated by Theodor Boveri (1891) (Fig. 5.2).

In the male, the original germ cells are called spermatogonia. They divide by mitosis for many generations. Just prior to maturation, the spermatogonia grow in size, their chromosomal material replicates, and the chromosomes associate in pairs. The cells are now known as primary spermatocytes. Each primary spermatocyte divides by meiosis into four spermatids which develop into spermatozoa without further cell division.

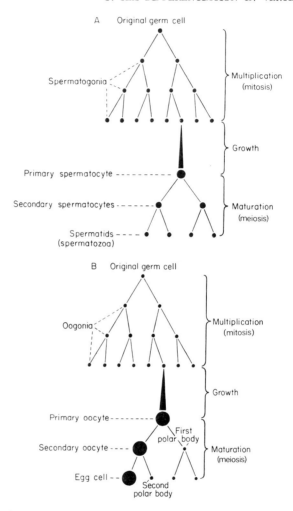

***Fig. 5.2.*** (A) Spermatogenesis and (B) oogenesis. Note that a primary spermatocyte gives rise to four haploid spermatozoa while a primary oocyte gives rise to a single haploid ovum. (After Boveri, 1891.)

The original germ cells of females are known as oogonia. They, too, divide by mitosis, and following the last mitosis, the oogonia grow in size, duplicate their chromosomal material, and pairing of homologous chromosomes takes place. The cells are now known as primary oocytes; and they characteristically remain in this stage for a long time, which may extend to many years (Peters, 1970). The remaining stages of meiosis occur during ovulation and fertilization. The first meiotic division results in a large secondary oocyte and a small primary polar body (Fig. 5.2). The second meiotic division occurs in different species either before, or after, fertilization. It is again an unequal division which gives rise to a large egg nucleus and a small second polar body. The first polar body may or may not undergo the second meiotic division to form two additional polar bodies. All the polar bodies disintegrate.

Thus, a spermatogonium undergoing meiosis gives rise to four spermatozoa, while an oogonium forms only a single egg cell. The numerical discrepancy between the two types of cells is, however, much greater than this. Spermatogonia continue to divide throughout the reproductive life of a male mammal. In man, several hundred million spermatozoa are used up in a single ejaculation. By contrast, cell division in human oogonia occurs only during fetal life. At the time of birth, the oogonia have already been transformed into primary oocytes (Austin, 1961), and they may remain in this stage for many years. Estimates of the number of primary oocytes present in the human female vary. Originally there may be several million (Witschi, 1956), but the number decreases both during and after embryonic life. In any case, only a small fraction of oocytes complete the maturation process to form egg cells. During the entire lifetime of a woman this number is barely five hundred. They mature at the rate of one per month during the reproductive period.

We see that the mammalian egg and sperm represent two homologous cell types, which are formed on the same basic plan, but which differ widely in size, in numbers produced, and in the time required for their development.

## III. Mammalian Gonads

Egg and sperm cells are produced in the gonads. The testes, or male gonads, are ovoid bodies, which, in man, reach a length of 4 to 5 cm when fully grown. In postnatal life, the testes occupy a position outside the abdomen, in the scrotum.

The testes are covered by the tunica albuginea. Inside there

is a convoluted network of seminiferous tubules (Fig. 5.3) in which
the sperm are produced. The walls of the seminiferous tubules
consist of many layers of cells, representing spermatogonia and spermatocytes. The innermost cells are spermatids and spermatozoa,
which are shed into the lumen of the tubules. A complicated network of ducts convey the spermatozoa into the vasa deferentia
and, finally, to the exterior.

In addition to the seminiferous tubules the testes also contain
interstitial cells, which produce male sex hormones, or androgens.

The ovaries, also, are elongated bodies which are somewhat
smaller than the testes. The ovary, too, has a dual function, namely,
to form ova and to produce sex hormones. Among the hormones
produced by the ovary, estrogen and progesterone are involved

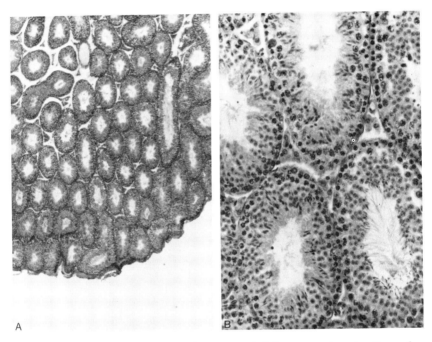

A                                    B

*Fig. 5.3.* Histology of mammalian testis (adult mouse—hematoxylin and
eosin stain). (A) Transverse section of part of testis, showing seminiferous
tubules ($\times$ 60); (B) as above at higher magnification, showing development of
spermatozoa ($\times$ 275). (Photographs contributed by M. Buehr.)

in controlling the secondary sexual characters. Each ovary is covered with a single layer of germinal epithelium, which surrounds the connective tissue, or stroma. Prior to the formation of ova, cluster of cells from the germinal epithelium invade the stroma. Each cluster forms a primary follicle (Fig. 5.4). The innermost cell is the primary oocyte, which on maturation will become an ovum, and this is surrounded by follicular epithelium cells. If the follicle is destined to mature, the follicular cells proliferate and a fluid, the liquor folliculi, develops inside the follicle. As a result of all these changes the follicle becomes so large that it bulges from the surface of the ovary and finally bursts. Both maturation and rupture of follicles are controlled by secretions from the pituitary gland. As a result, the ovum (strictly speaking the secondary

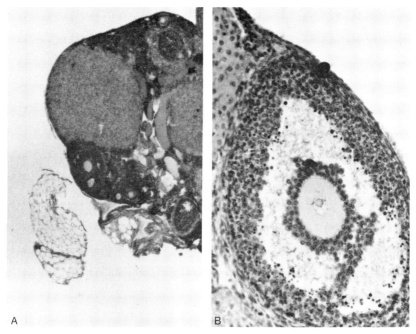

A                          B

*Fig. 5.4.* Histology of mammalian ovary (adult mouse—hematoxylin and eosin stain). (A) Part of ovary showing large corpora lutea and follicles in various stages of development (× 60); (B) developing follicle with oocyte at center (× 275). (Photographs contributed by M. Buehr.)

oocyte, since it has yet to undergo the second meiotic division) is released from the ovary and enters the oviduct, while the rest of the follicle forms the corpus luteum, which secretes the hormone progesterone. This hormone is concerned with the formation of the endometrium, which lines the uterus (Young 1961).

The oviducts are paired structures leading from each ovary to the uterus. The egg's journey through the oviduct takes 5 days and during this time it may meet with one of two fates. If fertilization does not take place, the corpus luteum disintegrates and the ovum together with the entire endometrium is discarded, resulting in menstruation. If, on the other hand, the egg encounters spermatozoa and is fertilized, the corpus luteum persists and the fertilized egg or zygote, after completing the second meiotic division and extruding the second polar body, undergoes a series of mitoses which transforms it into a blastocyst. This becomes embedded in the wall of the uterus where the subsequent development of the embryo takes place.

In nonhuman mammals, both the architecture of the female reproductive tract as well as the process of fertilization are basically similar to that in man, but the timing of ovulation and fertilization vary in different species.

## IV. Sex Differentiation in Other Vertebrates

In other vertebrates, the distinction between the two sexes is not quite as rigid as in mammals. In female birds, the gonads show considerable asymmetry and, as a rule, only the left one becomes a functional ovary. The right gonad tends to remain in a relatively undifferentiated state but has a distinct tendency toward testis formation (Lillie, 1952; Witschi, 1956; van Tienhoven, 1961). Further development is usually retarded by the presence of the left ovary, but if this is removed or severely damaged, growth is resumed in the right gonad, which may develop into a sterile testis, a fertile testis, a hermaphrodite gonad, or into an ovary. Development in a testicular direction tends to be accompanied by the appearance of male secondary sexual characters, particularly the growth of a comb.

In amphibians, hermaphroditism occurs at a relatively high rate. Differences in temperature as well as supermaturation of eggs can influence sex differentiation (Witschi, 1942). In some strains of *Rana temporaria* all animals spend the first 2 years of their lives as females, after which about one-half of them turn into males. In frogs, the production of sperm and eggs from the same gonad, or ovotestis, appear to be possible, and by artificial insemination offspring have been produced in this way.

Hermaphroditism is relatively common in fish. The developing gonads of many species pass through a bisexual stage, during which spermatocytes and oocytes are present, so that there is, in fact, a transient juvenile hermaphroditism (Forbes, 1961; D'Ancona, 1950). Subsequently, the gonad becomes either an ovary or a testis, but occasionally, the hermaphrodite condition persists into adulthood. In addition, a second type of consecutive hermaphroditism may be distinguished in which a fish may first become sexually differentiated and then transforms into the other sex. In protandrous hermaphrodites, the male phase develops first and the female phase later, while protogynous hermaphrodites are first female and subsequently develop into males (Yamamoto, 1969).

In *Gonostoma gracilis,* a deep-sea luminescent fish, Kawaguchi and Marumo (1967) found that individuals less than 7 cm were mostly males, while those more than 9 cm were always female. Sex succession took place between these sizes, where hermaphrodites were found. The opposite situation was described by Liu (1944) in *Monopterus javanensis,* a fish belonging to the order of Symbranchii. Small specimens were mostly female and large ones male, while hermaphrodite gonads were found in specimens of intermediate size.

The transitional hermaphrodite phase may be brief, and Bacci (1965) has used the term "false gonochorism" to describe animals which present exclusively male or female gonads during the major part of their life cycle but which nevertheless have a short intermediate phase during which the germ cells of one sex give way to those of the other. A striking example is provided by *Coris julis,* which at one time was regarded as a separate species from *C. giofredi.* Both inhabit the same shores of the Mediterranean. *Coris julis* is brightly colored and larger than *C. giofredi,* and it was thought that *C. julis* specimens are generally male while *C. giofredi* specimens are usually female. Bacci found that the few male *C. giofredi* specimens were larger than the females, while the few *C. julis* female specimens were smaller than the *C. julis* males. Subsequently, the transitional condition was discovered and all stages assigned to *C. julis,* which was thus shown to be a protogynous hermaphrodite.

Another interesting example of protogynous sex reversal was described by Fishelson (1970) in a Red Sea fish, *Anthias equamipinnis.* Natural populations were found to consist of 80–90% of females and only a few males. When females were placed in an aquarium with one or two males for several months, they did not change either in appearance or behavior. However, when groups of 20 females were kept without a male, one of the females would change into a male after 2 weeks. If this male was removed, another one of the females would become a male, and this process continued until all the females had changed into males by the end of the year. If, however, a male was kept behind glass, none of the females would change. It seems, therefore, that in this species, the sight of a male would inhibit male differentiation in the female fish.

In spite of a large number of reports of sporadic hermaphroditism in fish, little is known about its incidence. According to Atz (1964), hermaphrodite individuals are rare except in those species in which the condition is normally present.

Hermaphroditism is known to be the normal condition in one species of fish, *Rivulus marmoratus,* a member of the Cyprinodontidae (egg-laying toothcarps). Harrington (1961, 1968, 1971; Harrington and Kallman, 1968) reported that specimens taken in the wild in Florida are hermaphrodites, which habitually lay selffertilized eggs. The resulting offspring develop into clones, the members of which exhibit histocompatibility within, and histoincompatibility between, clones. Male fish could be obtained in the laboratory by incubating eggs at low temperatures. In addition, secondary males sometimes develop from hermaphrodites, usually late in life, by involution of the ovarian component of the ovotestis. No females were found.

Sex determination in fish presents a number of features of special interest. A number of species are known to have chromosomal mechanisms of sex determination which, however, are still in a relatively undifferentiated state. This makes it possible to gain some understanding of the interrelationship of chromosomes and environment in the process of sex differentiation. In addition, many fish are kept and bred in aquariums, with the result that far more is known about their reproductive habits than that of other lower vertebrates.

Some examples of the genetics of sex determination in fish, of which detailed studies are available, will now be described.

## V. Sex Determination in Fish

The first evidence pointing to a genetic basis for sex determination in fish was provided by Aida (1921), who discovered sex-linked inheritance of a color factor in the medaka, *Oryzias* (formerly *Aplocheilus*) *latipes* (see below). In the years that followed, the subject became firmly established by Winge (1922,a,b, 1927, 1934; Winge and Ditlevson 1938, 1947, 1948), who used the guppy, *Lebistes reticulatus*, as their experimental animal (the new generic name, *Poecilia*, is being used by some authors). Winge's findings have recently been extended by Haskins *et al.* (1970).

The guppy is a member of the Poeciliidae, or live-bearing tooth-carps, and like all members of its family, shows a distinct sexual dimorphism; in the male, which is smaller than the female, the anal fin is modified into a gonopodium (see Fig. 5.6).

While the females tend to be of a dull brownish or grey color, the males are brightly colored. The Y chromosome always contains color factors, and, in addition, the X chromosome may also contain color factors. However, most X-borne color factors are not expressed when present in females and show up only in males. By virtue of the color factors on the Y chromosome, the sex chromosome constitution of *Lebistes* (and other species of fish—see below) can be assessed with a high degree of accuracy, even though the sex chromosomes are not distinguishable in cytological preparations. The color factors have been called "genes" or "supergenes," but the more noncommittal term "factor" is preferred here.

Color factors on the Y chromosome of *Lebistes* include *maculatus*

($Y_{Ma}$) which is present in the "spot race" (Fig. 5.5) and results in
(1) a large black spot in the dorsal fin, (2) a large red side spot
below and in front of the dorsal fin, and (3) a black dot at the anus
(sometimes invisible); *armatus* ($Y_{Ar}$), which causes (1) two or three
red patches on the side, (2) one or two black spots on the side
in front of the dorsal fin, (3) a little black speck in the caudal fin,
and (4) a long, dagger-shaped, sulfur-colored prolongation of the
lower edge of the caudal fin; and the *pauper* factor ($P_{Pa}$), so-called
because of a factor on the Y chromosome relatively poor in decora-
tive value. *Pauper* produces a small red patch on the hind part
of the tail, a black speck set behind this patch, and, occasionally, a
black anal speck.

One of the original factors placed on the X chromosome, *sulfureus*,
was subsequently separated into a linked complex *coccineus* and
*vitellinus* ($X_{Co,Vi}$). Of these, *coccineus* causes a red color in the
lower edge of the caudal fin as well as a sulfur yellow color in the
caudal fin; while *vitellinus* is responsible for (1) sulfur yellow color
in the dorsal fin, (2) sulfur yellow color in the tail, (3) a small red

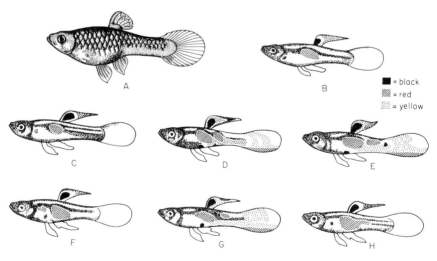

**Fig. 5.5.** X- and Y-borne color factors in *Lebistes reticulatus*. (A) ♀ $X_{Co,Vi}$,
$X_0$; (B) ♂ $X_0Y_{Ma}$ (spot race); (C)–(H) 6 sons; (C), (F), (H) $X_0Y_{Ma}$; (D),
(E), (G) $X_{Co,Vi}Y_{Ma}$. (After Winge, 1922.)

spot on the side of the tail, and (4) a dark side speck at the base
of the caudal fin.

Of eighteen color factors described by Winge (1927), nine were
located on the Y chromosome, three were on the X chromosome,
and five could apparently cross over between the X and the Y
chromosome. Only one color factor was autosomally determined.
None of the X-borne factors expressed the color effects in female
fish, although they showed the usual type of sex linkage. Subsequently, Dzwillo (1959) described a recent color variant, which
proved to be due to a factor located on the X chromosome with
a dominant effect in both males and females. This factor, known
as *half-black* ($X_{Bh}$), causes a heavy deposition of melanism in
the posterior part of the body.

In Winge's material, an occasional female would develop a
gonopodium and at the same time the color factors on its X chromosomes would become apparent. As a result of crossing such a female
to a $Y_{Ma}$ male, three males were obtained which did not show
*maculatus* and which, on breeding, had all female offspring. These
males, therefore, were almost certainly XX. After repeated cross-
ings, a male appeared which, when mated to his mother, produced
both male and female offspring. Thus, a race of *Lebistes* had been
developed in which both sexes were XX and sex was determined
by factors other than those located in the sex chromosomes. The
proportion of males and females in this strain was, however, rather
variable.

In another experiment, a few XY females carrying the *maculatus*
factor appeared and when mated to an $XY_{Pa}$ male gave rise to
XX, XY, and YY offspring in a 1:2:1 ratio. The YY males appeared
to be fully viable and fertile, although they give rise to only male
offspring. By contrast, YY males which were homozygous for *maculatus* could not be obtained and the existence of a recessive lethal
gene, closely linked to *maculatus*, was postulated (Winge and
Ditlevsen, 1938).

These findings have recently been confirmed by Haskins *et al.*
(1970) who obtained XY females by treating newborn XY fish,
carrying either *maculatus, armatus,* or *pauper,* with estradiol. These
sex-reversed females gave rise to YY males, which when mated
to XX females, had XY and YY offspring, which as a result of

further estrone treatment developed into females. When mated to YY males, they gave rise to all male offspring of XY and YY sex chromosome constitution. Analysis of the YY males showed an almost complete deficiency of fish that were homozygous for any of the three color factors, and the authors postulated the existence of recessive lethal genes, closely linked to each of the three color genes. However, YY males which were heterozygous for two color factors appeared to be fully viable and fertile.

In the medaka, Oryzias (formerly Aplocheilus) latipes, an egg-laying fish belonging to the Cyprinodontae, a red color factor situated on the sex chromosomes was discovered by Aida (1921, 1936). The inheritance of this factor suggested a sex chromosome constitution of XX in the female and XY in the male. Exceptional XY females occurred, however, and these produced the expected XX, XY, and YY offspring.

Yamamoto (1953–1969) described the experimental sex reversal of the medaka by means of oral administration of mammalian steroids. Estrogens, either estrone or stilbestrol, given to newly hatched and young fish, caused chromosomal males to develop into females, which subsequently produced YY males among their offspring. These were viable provided they were not homozygous for the red color factor, R. By subjecting the YY fish to further estrogen treatment, three YY females were obtained which produced all male offspring. Yamamoto and Matsuda (1963) have shown that only estrogenic steroids were effective as female inducers, nonsex hormone steroids having no effect.

Yamamoto (1958) also succeeded in producing XX males by the administration of methyltestosterone. Androsterone and testostrone propionate were less effective in their male inducing activity (Yamamoto et al., 1968).

In the platyfish, Xiphophorus (formerly Platypoecilus) maculatus, the formation of a gonopodium was induced in females as well as in castrated males by the addition of methyltestosterone to the aquarium water (Grobstein, 1948). Subsequently, Anders et al. (1969a,b) achieved complete sex reversal of chromosomal XY males into functional females by means of X irradiation. When pregnant females were exposed to doses of X rays between 1000 and 2500 R, a proportion of the XY offspring developed into females.

These proved to be fertile and when mated to normal XY males produced offspring consisting of XX females and XY and YY males in a 1:2:1 ratio. The sex of YY fish could again be reversed by X irradiation, but YY females were not fertile.

Both the X and the Y chromosomes of *Xiphophorus maculatus* contain a number of color factors and these, as well as the sex-determining mechanism, have been described by Gordon (1927–1959) and by Kallman (1965, 1970). Originally, Bellamy (1922) reported female heterogamy in aquarium specimens of this species but Gordon (1947, 1952) showed that both male and female heterogamety occurred side by side within the same species according to locality. In wild strains obtained from Mexico, males are XY and females XX, whereas the opposite situation occurs in specimens caught in the Belize River in British Honduras as well as in "domesticated" aquarium specimens. It is thought for this reason that the aquarium strains of platyfish originated from British Honduras.

All matings between males and females of different populations are fertile and produce offspring of normal sex, but not always in a 1:1 ratio. In particular, matings between Mexican females and males from British Honduras result in all male progeny (Chavin and Gordon, 1951), and from this it was concluded that the sex chromosome constitution of the males from British Honduras must be YY and that of the females WY. Although this interpretation of the identity of the sex-determining regions of the heterogametic sex chromosome in the Mexican population and the homogametic sex chromosome of the population is very likely correct, a less committed terminology would seem to be preferable, and in this text the sex chromosome constitutions of platyfish from Mexico and British Honduras will be designated as follows:

$$X_{Me}Y_{Me}, \; \male \; ; \; X_{Me}X_{Me}, \; \female \; ; \; X_{BH}X_{BH}, \; \male \; ; \; X_{BH}Y_{BH}, \; \female$$

The $X_{Me}X_{BH}$ individuals are males, while $X_{Me}Y_{BH}$ and $Y_{Me}Y_{BH}$ fish are female. Therefore, as already mentioned, Mexican females by British Honduras males give rise to all male offspring, while an $F_1$ female of constitution $X_M Y_{BH}$ mated to a Mexican $X_{Me}Y_{Me}$ male will produce a ratio of three females to one male. The other four matings give rise to equal numbers of males and females.

Kallman (1965) studied populations in Guatemala, which is situated between Mexico and British Honduras, and found fish with male and female heterogamety coexisting even within the same lake. Breeding experiments with females that were gravid when collected showed that they had been inseminated by either type of male. However, certain population crosses yielded some exceptional $X_{Me}Y_{BH}$ males (Kallman, 1968). This suggests that, although the $Y_{BH}$ chromosome has a fairly strong female-determining power, it evidently cannot impose female development under all conditions.

Closely linked to the sex differentiating segments of the sex chromosomes are a number of color factors, which have been widely used as markers of the sex chromosome constitution. Gordon (1948) described five such color factors, which give rise to characteristic spotting patterns, in "wild" or Mexican *Xiphophorus maculatus;* all five behave like members of an allelic series (Fig. 5.6). The factor *spot-sided* (also called *spotted-Sp*) causes large pigment cells, known as macromelanophores, to be scattered irregularly over the body; in *spotted belly* (*Sb*), pigment cells are concentrated along the midventral line; the *spotted-dorsal* (*Sd*) factor controls the development of one, two, or sometimes three small groups of macromelanophores in the dorsal fin; in *black-sided* or *nigra* (*N*), the macromelanophores form strong black bands; while in the *stripe-sided* (*Sr*) factor there are a fairly evenly produced series of macromelanophores, which produce horizontal lines of fine black spots on both sides of the body (Fig. 5.7).

Although the platyfish, *Xiphophorus maculatus*, has both sex chromosomes and color factors located on the sex chromosomes, the swordtail, *X. helleri*, has neither. In this species the sex ratios may fluctuate, and it has been suggested that sex determination has a polygenic basis (Kosswig, 1964). The chromosome pair of *X. helleri*, which is homologous to the sex chromosomes of *X. maculatus*, may be designated as xx in both sexes (Kosswig, 1932; Gordon, 1948; Anders and Anders, 1963). The two species will produce hybrids in the aquarium. The proportion of male to female offspring from the cross *X. maculatus* ♀ ($X_M . X_{Me}$) by *X. helleri* ♂ (xx) depends on the subspecies of *X. helleri* used in the experiment (Kosswig, 1931); in the reciprocal cross, *X. helleri* ♀ (xx) by *X. maculatus* ♂ ($X_{Me} Y_{Me}$), $X_{Me}x$ fish develop as females and most $xY_{Me}$ fish develop as males, although masculine differentiation occurs later than normal and a few $xY_{Me}$ become functional females. If $xY_{Me}$ males are backcrossed to X. *helleri* females, from 0 to 30% of fish become sexually differentiated into either males or females, depending on the *helleri* subspecies. The others begin to differentiate in a male direction, but remain in an intermediate stage.

Hybrids between platys and swordtails show instability not only in the process of sex determination but also in the disturbance of the development of the macromelanophores. Instead of forming well-defined patterns, the pigment cells of hybrids may continue to grow apparently without check, until eventually melanomas are formed, which may kill the fish. This phenomenon was discovered by Gordon (1927) and Kosswig (1927) and has since been intensively investigated (see Anders, 1967; Siciliano *et al.*, 1971).

Normally, the sex-linked color factors of X. *maculatus* determine the differentiation of macromelanophores, i.e., large pigment effector cells of lower vertebrates, with a diameter between 300 and 500 $\mu$m (Gordon, 1959). These originate from unpigmented melanoblasts, which arise in the neural crest and migrate to certain areas of the body, where they develop dendrites and the capacity to form melanin. These pigment-forming cells are now known as melanocytes. Finally, under the influence of a macromelanophore factor, the melanocytes become sessile, increase in size, and differentiate into melanophores and macromelanophores. However, in platyfish/swordtail hybrids, it seems that the melanocytes are prevented from differentiating into melanophores and instead undergo a greatly extended series of proliferations.

It is of interest that micromelanophore patterns, which are autosomally determined and in which the pigment cells remain smaller (100–300 $\mu$m), do not form melanomas in platyfish/swordtail hybrids.

The sex chromosomes of fish clearly present aspects of unusual interest. The assignment of a special pair of chromosomes for the purpose of sex determination would seem to be a luxury, rather than a necessity, in this class, since sex chromosomes may be present in one species and absent in a related one. This makes it possible to study sex chromosomes in an early evolutionary stage. Moreover, the presence of color factors on the sex chromosomes

A

B

C

D

E

F

*Fig. 5.6.* Inheritance of sex-linked color factors in platyfish, *Xiphophorus maculatus* from Mexico. (A) Female (undifferentiated anal fin) showing *spot-sided*, $X_{Sp}X_{Sp}$; (B) male (anal fin modified into gonopodium) showing *spotted-dorsal* and *stripe-sided*, $X_{Sd}Y_{Sr}$; (C) male offspring, $X_{Sp}Y_{Sr}$; (D) female offspring, $X_{Sp}X_{Sd}$; (E) female offspring, sex reversed by X irradiation, $X_{Sp}Y_{Sr}$; (F) male offspring of (E), $Y_{Sr}Y_{Sr}$. (Photographs contributed by F. and A. Anders.)

A

B

*Fig. 5.7.* Sex-linked color factor in *Xiphophorus maculatus* from British Honduras (female heterogamety). (A) Female $X_{BH}{}^N Y_{BH}$ ($Z^N W$); (B) male $X_{BH}{}^N X_{BH}{}^N$ ($Z^N Z^N$). (Photographs contributed by F. and A. Anders.)

of fish presents a tantalizing problem, which is further enhanced by the fact that the cells which are normally controlled by these color factors will grow excessively when chromosomes from differ-ent species come together in the same cell. The question of whether there is any direct connection between the color factors and the process of sex differentiation still remains unanswered. However, the fact that fish are lower vertebrates means that their development is far more open to the experimental approach than that of higher vertebrates, such as mammals, and there is thus a good chance that studies of the sex chromosomes of fish will yield pertinent information on the part played by sex chromosomes in the process of sex differentiation.

## VI. Induced Sex Reversal

The increasingly complex organization achieved by vertebrates, as we pass from fish to mammals, is also reflected in the degree of success achieved when attempts are made to channel the sexual development into a direction which is at variance with the genetic consitution.

Functional sex reversal in fish resulting from the oral administration of mammalian steroids was described in the last section. In the fighting fish, *Betta splendens,* Noble and Kumpf (1937) reported that the removal of the ovaries led to regeneration in 7 out of 100 fish; in all cases, the regenerated gonads were testes. Three of the secondary males fertilized eggs which gave rise to offspring of both sexes. When gonads were removed from male fish, regeneration always led to the development of testes.

Functional sex reversal has also been achieved in amphibians (see Burns, 1961; Foote, 1964). Chang and Witschi (1956) found that as a result of adding estradiol to the aquarium water containing young larvae of *Xenopus laevis,* all animals developed into females. Half of these produced only male offspring (unless treated with estradiol), indicating that the female is the heterogametic sex in this species (Fig. 5.8).

The time at which estradiol was effective in converting gonads into ovaries could be pinpointed with considerable accuracy. If the hormone was administered for 2 days starting at developmental stage 26 (Witschi, 1956), the tops of the gonads (i.e., the parts nearest the head) would differentiate into ovarian lobes, while the remaining parts developed into testicular tissue according to the

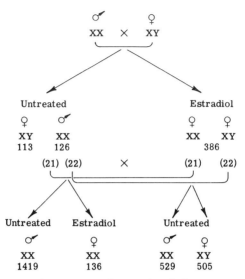

*Fig. 5.8* Sex reversal in *Xenopus laevis* (female heterogamety) by treatment with estradiol. (After Chang and Witschi, 1956.)

sex chromosome constitution. If, however, treatment was delayed until stage 27, only the lower part of the gonad (i.e., nearest the tail) differentiated into ovarian lobes. In order to achieve complete feminization, estradiol had to be administered for about 3 days. These results indicate that the response of the undifferentiated gonad to estradiol is limited to a specific period in development. On the other hand, the presence of androgen in the aquarium water had no effect on gonadal differentiation.

In contrast to these findings, parabiotic experiments carried out by Chang (1953), which involved the establishment of a common blood supply between male and female *Xenopus* larvae, resulted in the development of normal testes and greatly reduced ovaries. This suggests that once the gonads have differentiated, another mechanism operates by which secretions from the testes inhibit further ovarian development. Because of these contrasting findings, Chang and Witschi (1956) concluded that steroid hormones cannot be the primary sex inducers. This problem will be further discussed in the following section.

A number of attempts have been made to influence the sex differentiation of birds by treating embryos with steroid hormones (see Burns, 1961). Dantschakoff (1941) used a cross of chickens (*Gallus domesticus*) in which the segregation of marker genes resulted in the genetic sex being scorable at the time of hatching. Following the injection of folliculin into embryos, a proportion of genetic males developed into apparent females. None of these birds, however, laid eggs. Treatment with testosterone did not affect the sex of genetic females but merely resulted in various abnormalities.

In contrast to the result obtained with birds and amphibia, the administration of steroids to mammalian embryos has on the whole been unsuccessful in affecting the gonadal sex of the embryo (Burns, 1961; Wolff, 1962). The effect of extraneous sex hormones on early embryos is often lethal, and if they are administered later, the time at which sex reversal is possible may have been passed. Also, if the hormone is administered to the mother, it is not always known how much of it passes through the placenta.

The only clear-cut results on the effects of sex hormones on gonadal differentiation of a mammal have been obtained by Burns (1950, 1961) in a marsupial. In the American opossum, *Didelphis virginiana*, the young are born after a gestation time of only about $12\frac{3}{4}$ days (McGrady, 1938). Anatomically, however, the two sexes are indistinguishable for nearly another 2 weeks. In order to achieve any degree of sex reversal of the developing testis, a low dosage of estradiol dipropionate had to be begun soon after birth and treatment continued for 3 or 4 weeks. The low dose was necessary because of the lethality of the compound; even with 1 $\mu$g/day, a proportion of the treated animals died. In the survivors, the treatment resulted in testes being converted into ovotestes or even into "ovaries of remarkably normal histological structure," which did not, however, contain any germ cells. Burns (1950) interpreted these results on the assumption that estradiol dipropionate exerts a strong repression on the development of the embryonic testis and, if given early enough, ensures the survival of the germinal epithelium long after it has disappeared in an untreated testis. Subsequently, the germinal epithelium undergoes further proliferation and develops into an ovarian cortex. The lack of germ cells could be due to the fact that in the testis the germ cells normally disappear very early from the region of the germinal epithelium.

These results are remarkable in at least two ways. First, as mentioned above, they represent the only example of any degree of sex reversal achieved in mammalian gonads as a result of the administration of sex hormones. Second, the fact that estrogen has been effective in channeling the differentiation of male gonads into the female direction appears to be an exception to the rule that the sex hormone of the homogametic sex is not active in the sex differentiation of the embryo (Dantschakoff, 1941; also see following section).

A full evaluation of these findings must await further results on experimentally induced sex reversals. In the mean time, the possibility that sex differentiation in marsupials is relatively uncomplicated when compared with that of eutherian mamals should be kept in mind.

## VII. Embryological Basis of Sex Differentiation

In order to understand the process of sex differentiation, it is clearly necessary to know how the difference between males and females originally develops in the embryo.

The basic pattern of sex development is common throughout the vertebrates. Ovaries and testes originate from a pair of gonadal rudiments, while paired ducts are laid down through which the ripe sex cells will eventually be conveyed. Notwithstanding the great differences which these structures may exhibit in adult males and females, it is a striking fact that an embryo, whatever its sex chromosome constitution, originally develops the rudiments of both male and female sex organs. Immediately after being formed, the gonads pass through an apparently undifferentiated state and contain the forerunners of both ovarian and testicular tissue. In addition, each embryo starts off its sexual development with a pair of male (Wolffian) as well as a pair of female (Mullerian) ducts. The process of sex differentiation may thus be regarded as originating in the hermaphrodite condition and consisting of the progressive development of the organs of one sex at the expense of that of the other (Fig. 5.9).

In the mammalian embryo, the gonadal primordia arise as a

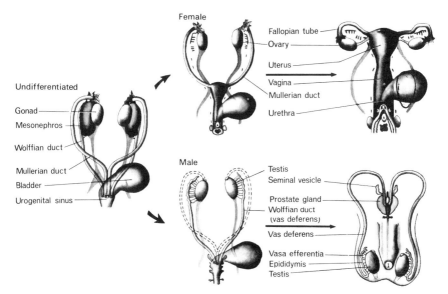

*Fig. 5.9.* Mammalian sex differentiation from potentially hermaphrodite rudiments (From A. Jost, *Sci. J.* 6, 67, 1970.)

pair of ridges, which grow out from the internal surface of the mesonephros, or embryonic kidney. These ridges represent only the somatic parts of the gonads. The primordial germ cells originate from an entirely different source, probably the endodermal epithelium of the yolk sac (Witschi, 1948, 1956; Burns, 1955). From their place of origin, the primordial germ cells migrate into the gonadal or genital ridges, where they settle and proliferate.

At first the gonadal primordia, including the germ cells, develop in an apparently identical fashion in both sexes. Two principal regions may be distinguished, an outer cortex and an inner medulla. The medulla consists of strands of the mesonephric blastema, while the cortex consists of thickened peritoneal epithelium. Between the cortex and medulla there is a thin layer of cells known as the albuginea.

The cortex is potentially capable of ovarian differentiation, while the medulla may develop into a testis. Normally, the development of the gonad follows one or other course according to the sex chromosome constitution. If the chromosomes are XX, the cortex develops while the medulla regresses and the gonad becomes an ovary; if the sex chromosomes are XY, the medulla develops at the expense of the cortex, resulting in the formation of a testis (Fig. 5.10). In mammals, the Y chromosome is the principal sex differentiator: rare exceptions aside, if a Y chromosome is present, the gonad becomes a testis, and in its absence, ovarian differentiation occurs (Chapter 6). The number of X chromosomes present is not important in this context.

In human embryos, a definite gonadal ridge is first seen about the fifth week of gestation (van Wagenen and Simpson, 1965). During the following week, the gonads continue to grow without showing any histological difference between males and females. The first signs of sexual differentiation of the gonads are seen in male embryos of about 7 weeks. At the periphery of the gonadal rudiments, the tunica albuginea begins to be formed by small

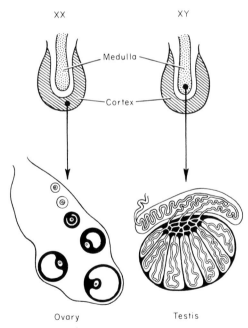

*Fig. 5.10.* Development of ovary and testis in mammals from potentially hermaphrodite gonadal rudiment. (From Mittwoch, 1967d.)

closely packed cells, and this area soon becomes depleted of germ cells. The cells together with supporting (Sertoli) cells aggregate in cordlike strands in the center of the gonads; these sex cords are the forerunners of the seminiferous tubules (Jost, 1970b; Gier and Marion, 1970).

Ovaries at this stage can be distinguished only by elimination, i.e., through the absence of testicular differentiation. The characteristic ovarian differentiation, which consists of the formation of follicles and the delineation of the ovarian stroma does not occur until much later. At first, indistinct cords are, in fact, formed in the medulla; these degenerate, however, and subsequently the cortex enlarges by what is often called a second proliferation. Gradually the cortex becomes the principal region. The delay in ovarian differentiation when compared with that of the testis is illustrated in Fig. 5.11 and Table 5.1.

A remarkable feature of ovarian development is that the prophase of the first meiotic division occurs very early in development. In human embryos, they are first seen at about 12 weeks of age, and the entire complement of primary oocytes is formed by the seventh month of pregnancy; in some mammals, e.g., the rabbit and the ferret, oocytes are formed during the neonatal period (Peters, 1970). Primary spermatocytes are not formed until shortly before puberty and their production continues throughout the reproductive life.

As mentioned before, the embryo is initially equipped with two

TABLE 5.1

CHRONOLOGY OF TESTICULAR AND OVARIAN DIFFERENTIATION IN THREE MAMMALIAN SPECIES[a]

| | Rat (days) | Rabbit (days) | Man (weeks) |
|---|---|---|---|
| Seminiferous tubules | 13–14 | 14–15 | 6–7 |
| Leydig cells | 16 | 19 | 8 |
| Primary oocytes | 18 | Postnatal | 11–12 |
| Ovarian follicles and stroma | Postnatal | Postnatal | 18–28 |

[a] From Jost, 1970a.

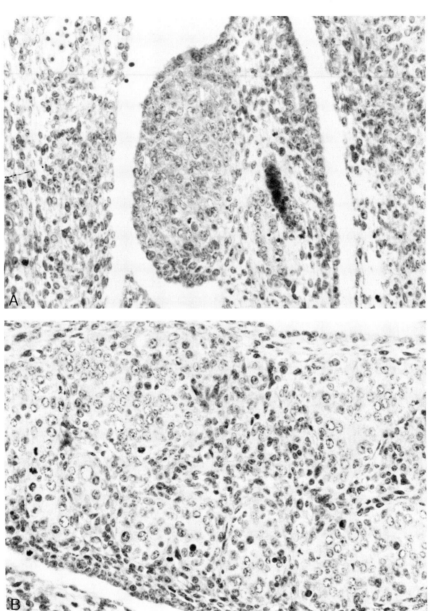

*Fig. 5.11.* Histology of differentiating testes and ovaries in embryos of the mouse. (A) Undifferentiated gonad, embryo 12th day of pregnancy ($\times$ 280); (B) differentiating testis, embryo 13th day of pregnancy, showing beginning of formation of seminiferous tubulus ($\times$ 280).

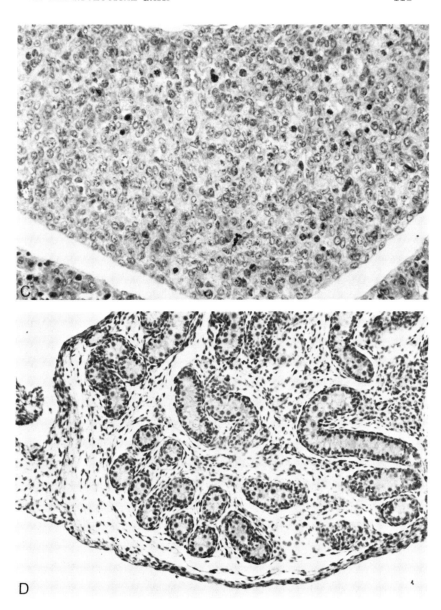

*Fig. 5.11.* (*continued*). (C) Gonad, female embryo 13th day of pregnancy (× 280); (D) testis, embryo 19th day of pregnancy (× 120).

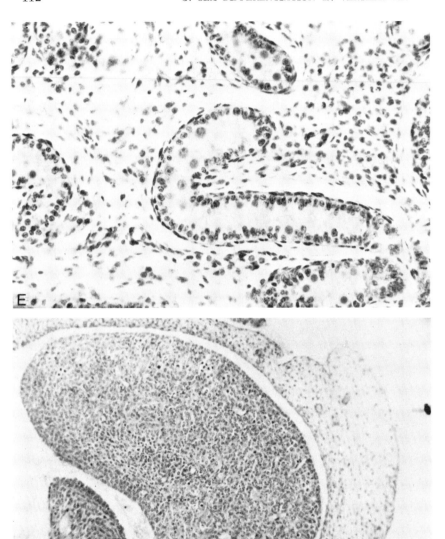

**Fig. 5.11.** (*continued*). (E) Same as (D) (× 280); (F) ovary, embryo 19th day of pregnancy (× 120).

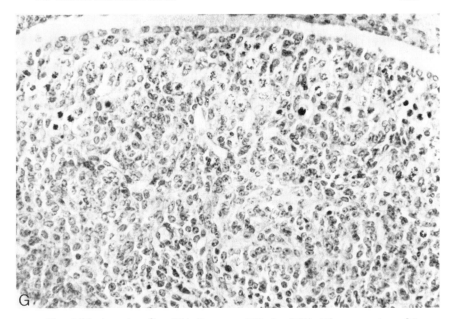

*Fig. 5.11.* (*continued*). (G) Same as (F) (× 280). The prococious differentiation of the testis is a typical feature of mammalian sex differentiation. (Photographs contributed by H. Peters.)

sets of gonadal ducts; in the human embryo, these persist side by side until the eighth week. The Wolffian duct originates as the excretory duct of the mesonephros, while the Mullerian duct grows from a funnel opening into the abdominal cavity at the cephalic end of the mesonephros. Both ducts join the urogenital sinus.

The first sign of somatic sex differentiation occurs in male embryos when the Mullerian duct begins to regress and then disappears. This is followed by the formation of buds of the future prostate gland. Each Wolffian duct develops into a vas deferens, which at its caudal end forms the seminal vesicles. The urethra terminates in the genital tubercle, which becomes the penis (Fig. 5.9).

In female embryos, the Wolffian ducts gradually disappear, while the Mullerian ducts persist and become the Fallopian tubes and the uterus. In humans, the ducts fuse at the base to form a single uterus, but in many mammals there are two uterine horns which fuse only at their caudal ends. The caudal portions of the Mullerian ducts form the primitive vagina.

The idea that hormones secreted by the embryonic testis may play a part in the differentiation of the male phenotype is over 50 years old. It was originally put forward by Tandler and Keller (1911) and by Lillie (1917) on the basis of observations on the freemartin, a sterile twin occurring in cattle (see Chapter 6). The authors showed that a freemartin results only if the twins are of unlike sex, and they postulated that the genetic female becomes transformed into an intersex by hormones secreted by the male twin. The male, however, is not affected by the female twin.

Although there has since been a great deal of discussion, and some controversy, on the subject of the causation of freemartins in cattle and other species (see Short, 1970; Ilberry and Williams, 1967; Matton-van Leuven and François, 1970; Benirschke, 1973) the theory that differentiation of the male phenotype is dependent on a hormone secreted by the embryonic testis is now well established. The evidence has been obtained particularly from castration experiments on early embryos. Jost (1947, 1965, 1970a) found that if the gonads were removed from rabbit embryos aged 19 days, i.e., after the testes have differentiated but while the genital tract is still in the indifferent stage, the embryos developed female characteristics, independently of their chromosomal sex (Fig. 5.12). In both sexes, which were scored in accordance with the histology of the excised gonads, the Wolffian ducts disappeared, while the Mullerian ducts developed into Fallopian tubes and uterine horns, and the external genitalia were of the female type. If male embryos were castrated 2 days later, the development of the Mullerian ducts were largely inhibited and there was some development of prostatic buds, but the external genitalia were still female. In order to ensure complete masculinization, the embryonic testes had to remain *in situ* until the 24th day, when the male organs had become firmly established. These results indicate that the embryonic testis promotes the development of the Wolffian duct and its derivatives and surpresses the development of the Mullerian duct, thus imposing the male phenotype on the embryo. Embryonic ovaries, on the other hand, are not required for the achievement of the female phenotype, which develops spontaneously in the absence of testes.

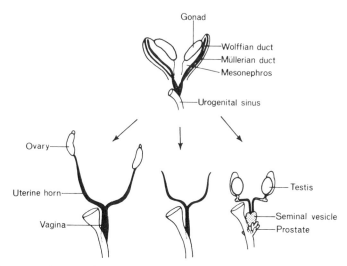

*Fig. 5.12.* Absence of gonads leads to female sex differentiation in mammals. From Jost, A. *Mem. Soc. Endocrinol,* **7,** 49–62, 1970).

Masculinization of female embryos by injecting androgens, such as testosterone, into the mother has been obtained in the hamster (Bruner and Witschi, 1946) and in cattle (Jost *et al.,* 1963). The treatment, however, did not inhibit the Mullerian duct derivatives in either species, and it has been postulated that the embryonic testis may secrete a second substance, of as yet unknown nature, which inhibits the Mullerian ducts (Jost, 1970c). However, the simple hypothesis, according to which the stabilization of the Wolffian duct and the regression of the Mullerian duct are under the control of a single hormone but depend on its concentration at given stages of development, still remains possible (Polani, 1970).

Experiments in which isolated reproductive tracts were kept in organ culture have confirmed the hormone-secreting capacity of the mammalian embryonic testis and its role in the maintenance of the Wolffian duct (Price and Ortiz, 1965; Price, 1970). The testes of guinea pig embryos aged 22–23 days, when the gonads still appear undifferentiated histologically and when the Mullerian duct is only beginning to be formed, had a stimulating effect on the growth of rat prostate when they were grown together in organ culture. In guinea pig embryos aged 26–27 days, both Mullerian and Wolffian ducts are present in both sexes; at 29–30 days, the Mullerian ducts of male embryos and the Wolffian ducts of female embryos begin to regress. If male reproductive tracts from embryos aged 26–27 days were explanted, the Wolffian ducts were retained only in the presence of embryonic testes or another androgen-secreting source. On the other hand, the survival and development of Mullerian ducts of female embryos were independent of the presence of embryonic ovaries. In these experiments, the Mullerian ducts of male embryos regressed under all conditions, suggesting that they may have been conditioned to take this course before they were explanted (Price, 1970).

While the key role played by the embryonic testis of mammals in the differentiation of the male phenotype is now generally accepted, there is as yet no consensus of opinion of what determines the apparently bipotential gonad to develop into a testis. Evidence thus far is lacking for the existence of a hormone under whose influence the gonad might differentiate into a testis. The effect of the antiandrogen cyproterone acetate may be relevant in this connection. Cyproterone acetate causes the feminization of male embryos to an extent which varies somewhat between different species (Neumann et al., 1970). Treated male rats and mice had female external genitalia and a vagina, but persistent Wolffian ducts; whereas in rabbits and dogs, the Wolffian ducts had regressed. In all species, the gonads were testes, which suggests that testicular differentiation is not influenced either by androgens or by antiandrogens.

Sometime during the course of development the Y chromosome must clearly exert its own specific effect, and I have previously suggested that the effect of the mammalian Y chromosome may be to increase the number of mitotic divisions in the cells of the gonadal rudiments (Mittwoch 1969, 1970). Measurement of gonadal volumes carried out independently by Lindh (1961) and by Mittwoch et al. (1969) have, indeed, shown that in embryonic littermates of rats and of golden hamsters (*Mesocricetus auratus*) incipient testes grow faster than incipient ovaries (Fig. 5.13; Table 5.2). The larger size of future testes was evident in embryos with

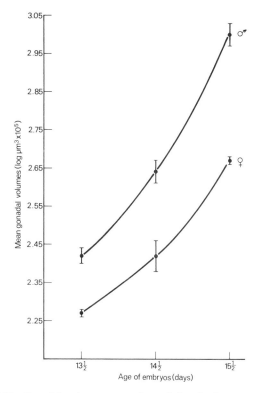

*Fig. 5.13.* Gonadal volumes in male and female littermates of rat embryos. Means and standard errors. (After Mittwoch *et al.,* 1969.)

male sex chromosomes before any histological differentiation of the gonadal rudiment was detectable. It is possible, therefore, that the mammalian testis may need to reach a certain size by a given stage in development in order to become a testis and that, if it fails to reach this size, the gonad will develop into an ovary (Mittwoch *et al.,* 1969). According to Jost (1970b) "some triggering mechanism seems to divert the male gonad from this slow ovarian evolution and to impose a more rapid and early testicular differentiation."

The possibility that the Y chromosome itself may be the triggering mechanism is suggested by three sets of data. (1) In *Sorghum* plants, the presence of supernumerary B chromosomes has been

TABLE 5.2

GONADAL VOLUMES IN MALE AND FEMALE RAT EMBRYOS[a]

| Age of embryos (days) | Gonadal volumes ($\mu m^3 \times 10^5$) | |
| --- | --- | --- |
| | Males | Females |
| $13\frac{1}{2}$ | 285 | 160 |
| | 281 | 196 |
| | 236 | 189 |
| | 250 | 179 |
| | 260 | 200 |
| | | 200 |
| | | 183 |
| Means | 262 | 187 |
| $14\frac{1}{2}$ | 436 | 250 |
| | 390 | 241 |
| | 489 | 284 |
| | | 220 |
| | | 349 |
| | | 347 |
| | | 189 |
| Means | 438 | 269 |
| $15\frac{1}{2}$ | 730 | 472 |
| | 941 | 454 |
| | 873 | |
| | 948 | |
| | 1166 | |
| | 1208 | |
| | 1017 | |
| Means | 983 | 463 |

[a] From Mittwoch et al., 1969.

found to increase the number of mitoses in pollen grains (Darlington and Thomas, 1941); (2) in several species of *Drosophila*, the Y chromosome forms evanescent lampbrushlike loops, which are active in RNA synthesis, in the spermatocytes of male larvae (Hess, 1970). This suggests the possibility that the mammalian Y chromosome may also have periods of intense RNA synthesis, which may be confined to cells of the gonadal rudiment and which stimulates additional mitotic divisions (Mittwoch *et al.*, 1969). (3) There is now a considerable body of evidence suggesting an interrelation between mitotic divisions and cell differentiation (Tsaney and Sendov, 1971; Bellairs, 1971; see Chapter 7). In Chapter 6, transdetermination experiments carried out by Hadorn and collaborators (Hadorn, 1968; Wildermuth, 1970) will be referred to. Essentially, the authors found that if cells of the imaginal discs of *Drosophila*, which are determined to give rise to specific organs, are kept in culture for several generations, they would sometimes give rise to organs other than those for which they were determined. This transdetermination took place only following a period of rapid cell division. By analogy, we may regard the mammalian gonad to be determined to become an ovary unless a Y chromosome is present, when additional cell divisions will bring about transdetermination leading to the development of a testis.

When we turn to the problem of sex determination in birds, the first fact to be noticed is that here the female is the heterogametic sex, having XY (ZW) sex chromosomes, while the male is XX (ZZ) (Fig. 5.14). The situation, therefore, appears to be precisely the opposite to that found in mammals, though it must be admitted that at this time there is no clear evidence as to the sex-determining role of the individual sex chromosomes in birds. This is because it is not yet known what the sexual development of otherwise diploid birds with XXY or XO sex chromosomes would be. Triploid chickens with XXY chromosomes present an intersexual appearance; the exact internal and external morphology still require elucidation (Bloom, 1971). It is likely that the triploid chicken with a left ovotestis described by Ohno *et al.* (1963) had XXY (ZZW) sex chromosomes. While the effects of sex chromosome aneuploidy on the sexual development remains to be worked out, the possibility that the avian Y chromosome may play a part in ovarian differentiation must be borne in mind.

Although the development of the gonads in bird embryos is basically similar to that in mammals, there are some striking differences in detail (Lillie, 1952; Witschi, 1956; van Tienhoven, 1961; Romanoff, 1960). The gonads of chick (*Gallus domesticus*) embryos like those of mammals are laid down on the median surface of the mesonephros, which thus become the genital ridge. Subsequently, the primordial germ cells migrate from the yolk sac

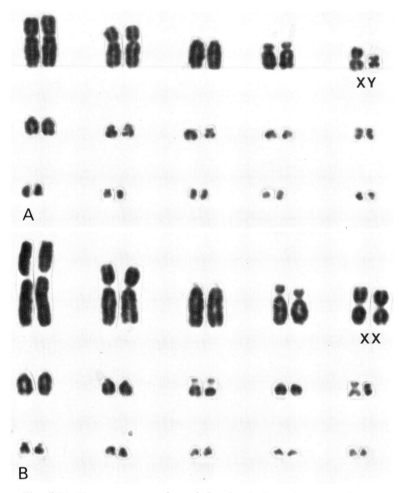

**Fig. 5.14.** Karyotypes in male and female chickens (some of the micro-chromosomes have been omitted). (A) Female, (B) male.

epithelium and settle in the genital ridge. Originally, the gonads appear indistinguishable, and all embryos have two pairs of ducts, i.e., two mesonephric or Wolffian ducts as well as two Mullerian or oviducts. A characteristic feature in the development of the sex organs in bird embryos is the marked asymmetry between the left and the right side. With few exceptions, only the left gonad of female birds develops into a functional ovary, while the right one remains in a rudimentary state. However, in younger embryos, gonadal asymmetry, with a larger left and a smaller right gonad, occurs in both sexes (Swift, 1915). During this so-called indifferent period, the left gonad of both sexes contain a well-developed germinal epithelium (Fig. 5.15), which is actually an incipient ovarian cortex, capable of giving rise to a second series of sexual cords and a true cortex if stimulated by female sex hormones (Lillie, 1952). In male embryos, the traces of germinal epithelium persists until the eleventh day of incubation. The right gonad contains little if any germinal epithelium and can usually differentiate only into a testis. If the right ovary of female birds is removed or diseased the left gonad will enlarge and differentiate into testicular tissue. The left gonad of both male and female bird embryos, therefore, is at first potentially hermaphroditic, while the right gonad has only the potential for male differentiation.

Measurements of gonadal volumes in male and female chick embryos, as established by chromosome analysis (Mittwoch *et al.*, 1971), have confirmed the marked asymmetry in size between left and right gonads in embryos until roughly the seventh day of incubation; during this time, the gonads of male embryos appeared to be larger than the corresponding gonads of female embryos (Table 5.3). However, from about the eighth day onward, the left gonad of female embryos became larger than that of male embryos (Fig. 5.16). It is known that from the ninth to the eleventh day of incubation, there is a rapid second proliferation of the germinal epithelium accompanied by a rapid increase of the primordial germ cells contained in it in the left gonads of female embryos (Lillie, 1952). During the same time, the germinal epithelium in the left gonads of male embryos can be seen to regress (Fig. 5.15). It appears, therefore, that ovarian differentiation in chick embryos is accompanied by a growth rate which is greater than that of the gonads of male embryos; this, in turn, suggests that, in line with their different sex chromosome constitution, the process of gonadal differentiation in birds may be the antithesis of that in mammals. We have suggested, therefore, that the left gonad of bird embryos may be required to reach a certain size by a given

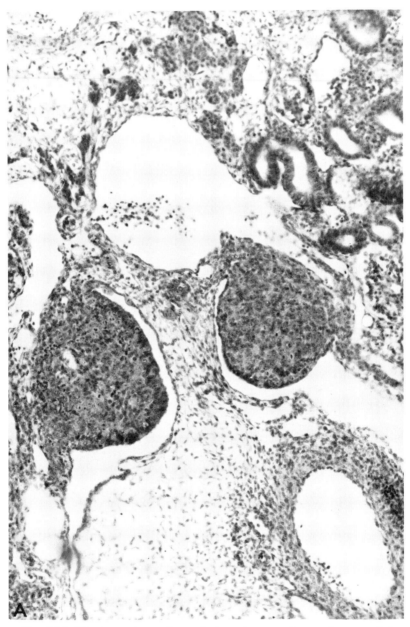

*Fig. 5.15.* Histology of gonads in chick embryos. Left gonads are on left (A)–(D) or bottom (E)–(H) of photographs (×350). Note thick germinal epithelium in left gonads of both sexes. (From Mittwoch *et al.,* 1971.) (A) Female, 6 days.

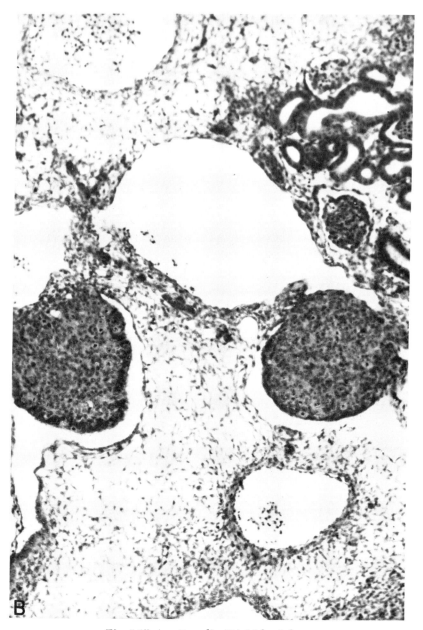

*Fig. 5.15. (continued).* (B) Male, 6 days.

*Fig. 5.15.* (*continued*). (C) Female, 7 days.

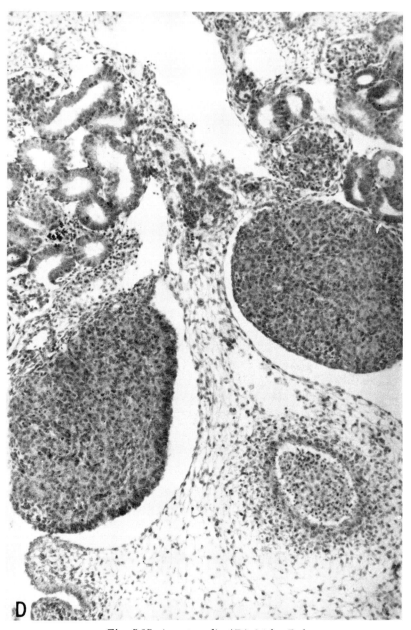

*Fig. 5.15.* (*continued*). (D) Male, 7 days.

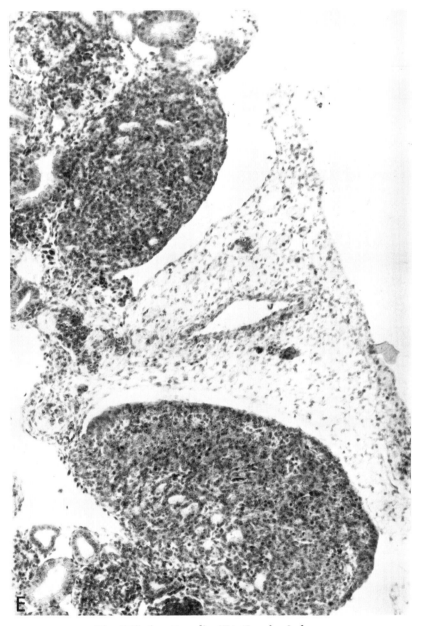

*Fig. 5.15.* (*continued*). (E) Female, 8 days.

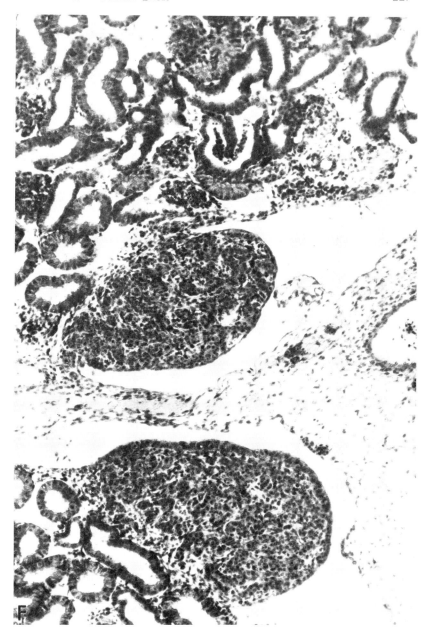

Fig. 5.15. (continued). (F) Male, 8 days.

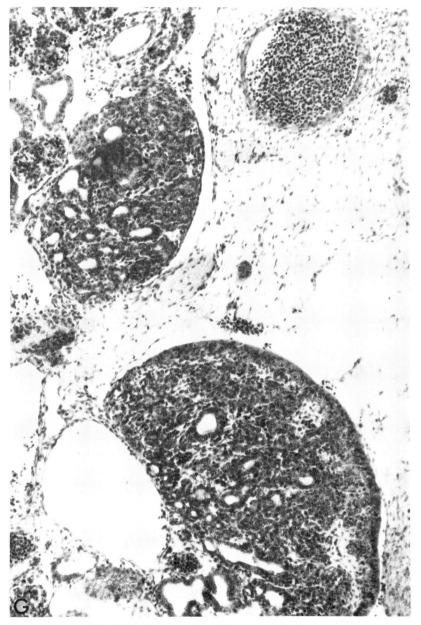

*Fig. 5.15. (continued).* (G) Female, 9 days.

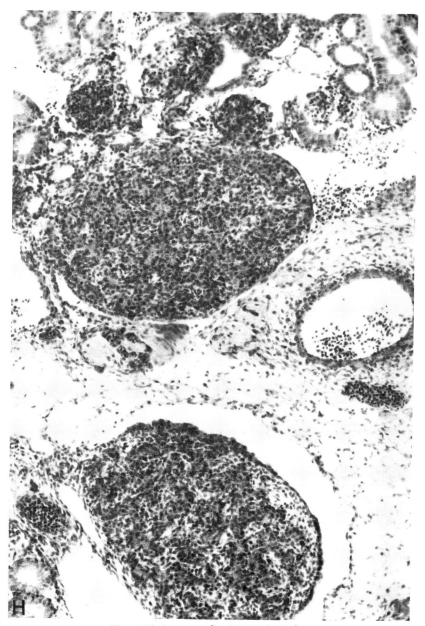

*Fig. 5.15.* (*continued*). (H) Male, 9 days.

TABLE 5.3
Gonadal Volumes in Chick Embryos[a,b]

| Age (days) | Males | | | | | | | Females | | | | | | |
|---|---|---|---|---|---|---|---|---|---|---|---|---|---|---|
| | | Left | | Right | | Left/right | | | Left | | Right | | Left/right | |
| | n | Mean | SE | Mean | SE | Mean | SE | n | Mean | SE | Mean | SE | Mean | SE |
| 5 | 2 | 158 | 20.0 | 74 | 6.0 | 2.12 | 0.09 | 3 | 107 | 22.5 | 62 | 15.0 | 1.74 | 0.08 |
| 6 | 3 | 403 | 44.0 | 282 | 45 | 1.46 | 0.09 | 1 | 289 | — | 159 | — | 1.82 | — |
| 7 | 2 | 924 | 21.5 | 547 | 85 | 1.72 | 0.23 | 4 | 665 | 83 | 398 | 76 | 1.72 | 0.10 |
| 8 | 3 | 987 | 133 | 794 | 169 | 1.29 | 0.10 | 8 | 1050 | 51 | 470 | 36 | 2.18 | 0.13 |
| 8.75 | 6 | 1491 | 158 | 1129 | 138 | 1.35 | 0.07 | 4 | 1678 | 107 | 590 | 76 | 2.95 | 0.30 |
| 9 | 4 | 1299 | 140 | 1003 | 117 | 1.32 | 0.10 | 3 | 2010 | 326 | 767 | 46 | 2.62 | 0.37 |
| 10 | 6 | 2059 | 247 | 1591 | 226 | 1.32 | 0.09 | 5 | 2634 | 425 | 768 | 70 | 3.41 | 0.37 |

[a] Measured as $\mu m^3 \times 10^5$. $n$ is the number of individuals.
[b] From Mittwoch et al., 1971.

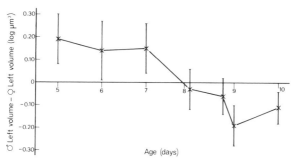

**Fig. 5.16.** Mean gonadal difference (log scale) and standard errors between left gonads of male and female chick embryos. (From Mittwoch *et al.,* 1971.)

stage of development in order to develop into an ovary and that if it fails to reach this size, the gonad will develop into a testis or an ovotestis (Mittwoch *et al.,* 1971).

Furthermore, castration experiments on bird embryos carried out by Wolff and Wolff (1951) have shown that in birds the ovary is the dominant embryonic gonad. In the course of normal development, both Mullerian ducts of male embryos regress, while in female bird embryos only the left Mullerian duct develops into an oviduct, while the right one regresses apart from a caudal remnant. When the gonads of chick and duck embryos were destroyed by X rays, the embryos developed a predominantly male phenotype with male-type syrinx and penis, apart from the persistence of both Mullerian ducts. It seems, therefore, that the embryonic testes of birds cause the Mullerian ducts to regress, whereas the major part of sex differentiation devolves on the embryonic ovary.

It is possible that the differences in the mechanism of gonadal differentiation between birds and mammals may be connected with differences in their productive physiology (Mittwoch, 1971). The mammalian embryo develops inside the mother, and it is known that a certain amount of female sex hormones will traverse the placenta and enter the embryo (Diczfalusy and Mancuso, 1969), so that embryos of both sexes are exposed to them. This means that differentiation of the reproductive tract by means of female sex hormones would not be a feasible mechanism in mammals. Accordingly, in mammalian embryos, there are no female sex hor-

mones originating from the embryonic ovaries, but male development is dependent on male sex hormones secreted by the gonads
of male embryos. If a Y chromosome does, indeed, speed up the
growth and development of the gonadal rudiments, it may need
to be present in the male during the period of its competitive
struggle for hormonal supremacy. By contrast, the avian embryo
is removed from its mother's sphere of influence long before its
sexual development is threatened. The embryonic testes are required only for the relatively minor task of causing the Mullerian
ducts to regress, while the major task of embryonic sex differentiation devolves on the ovary, whose extra growth at this stage might
be advantageous in preparing it for its ultimately more demanding
role of egg production.

This scheme must at present be regarded as speculative. In particular, more data are required regarding mechanisms of sex differentiation in viviparous nonmammalian vertebrates such as, for
instance, in certain snakes. Howover, even with the limited data
available in birds and mammals, it is becoming apparent that the
difference in sex chromosome mechanisms, i.e., female heterogamety
in one class and male heterogamety in the other, are associated with
some clear-cut differences in the relative growth rates of male and
female differentiating gonads. As far as they go, therefore, the
data support the hypothesis that sex chromosomes function by regulating the growth of the gonadal rudiments.


## VIII. Sex Determination in Three Invertebrates

It is clearly impractical at this point to attempt a review of the
mechanisms of sex differentiation found in invertebrate animals.
The subject has been reviewed by Bacci (1965). However, the findings on three widely separated species will be briefly described
in order to show that the relationship between sex differentiation
and growth is not confined to the mechanism of gonadal differentiation in vertebrates.

*Ophryotrocha puerilis,* a small polychaete worm, is a protandrous
hermaphrodite species, which passes from a male phase via a hermaphrodite phase to an adult female phase; this   is sometimes fol-

lowed by a final male phase (Bacci, 1965). It had already been shown by Hartmann and Huth (1936) that if the posterior segments were removed from female worms, cutting them down to five segments, the segments would regenerate and the worms revert to the male phase. After a further regeneration of segments, the worms turned once more into females. Two subspecies could be distinguished which differed with regard to the size attained by individuals when oocytes were first formed. In individuals caught in the Mediterranean Sea at Naples, small oocytes first appeared at a mean size of eighteen chaetigerous segments, compared with twenty chaetigerous segments in specimens originating from the English Channel in Plymouth. However, in individuals from Naples, the differentiation of oocytes was completed, whereas this stage was seldom reached in Plymouth specimens, in whom reversion from female to male occurred more readily. These findings suggest that as a means of adaptation to the lower temperature in the English Channel, the Plymouth population had evolved a slower rate of development than the Italian population and that this difference persisted even when the two groups were placed in identical environments in the aquarium.

The relationship between sex differentiation and temperature is particularly well illustrated in the mosquito, *Aedes stimulans*. In several species of mosquitoes, male development occurs only if the larvae are reared at low temperatures. At higher temperatures, all larvae develop as females. Anderson (1967) found that the sex organs of normal males and females originate as a pair of imaginal discs, which in males arise in the ninth sternite, while in females one pair is formed in the eighth and one in the ninth sternite. For a while, the discs develop in the same way in both sexes. Divergent development begins in the discs of the ninth segments: In the normal males, evagination occurs only in the third instar, while in normal females the pairs coalesce to form a median evagination in the fourth instar. In male larvae, which are kept at elevated temperatures, three pairs of imaginal discs develop, one pair in the eighth sternite, as in females, and two pairs in the ninth sternite; in the fourth instar development according to the female pattern occurs.

Last, we may mention *Bonellia viridis*, a marine worm belonging to the order Echiuroidea. This species shows an extreme degree of sexual dimorphism. The female has a proboscis which, when extended, may reach a length of 80 cm, while the male measures only 1 to 3 mm; two or three males normally live parasitically on the proboscis or in the uterus of a female. This difference in development is brought about almost entirely by environmental factors.

It was shown by Baltzer (1914, 1937) that the larvae are sexually undifferentiated and most, if not all, are sexually bipotential. If a larva settles on the proboscis of a female, it will develop into a male, whereas larvae which settle some distance away from a female becomes females. However, a small proportion of larvae from late broods develop into males in the absence of a female influence. It also appears that sexual development can further be influenced by the amount of potassium in the water (Herbst, 1935).

In summary, there appears to be abundant evidence that the divergent sexual development of organisms is associated with differential rates of growth. In less highly developed organisms, these differences may be dependent on environmental differences, such as temperature or the chemical composition of the surrounding medium. However, as greater complexity is achieved, sex chromosomes are evolved and these take over the function of regulating the growth rate of the organs responsible for sexual development, thus making the process of sex differentiation increasingly independent of the environment.

## IX. Sex Factors in Bacteria

It should be clear from the preceding text that the process of sex differentiation, which culminates in the production of eggs and spermatozoa by different individual organisms, is a feature of the evolution of higher organisms. However, various types of cell fusion occur also in lower organisms and the term "sex" is frequently used in connection with such processes. The so-called "sex factor" in *Escherichia coli* is a well-known example and this topic will therefore be briefly considered.

The ability of *Escherichia coli* cells to give rise to progeny which

are recombinants of two parental strains was discovered by Lederberg and Tatum (1946). Hayes (1952) has shown that in this
type of conjugation, genetic material could be transferred only
from one strain of bacteria to another and not vice versa. Thus,
two types of bacteria came to be distinguished: donor bacteria
and recipient bacteria, which were thought to exhibit sexual differentiation (Jacob and Wollman, 1961). This, however, seems to con-
fuse fertilization with copulation. Gametic fusion in higher organisms involves two nuclei which are genetically equal and do not
show any polarity. Sex differentiation in higher organisms is seen
in the nongenetic parts of the gametes, as well as in the organisms
which produce them.

It is now known that the donor state in *Escherichia coli* is con-
ferred by an agent called F, or sex factor, which is readily transmitted to recipient cells independently of the bacterial chromosome
(Hayes, 1968). It is, therefore, unlikely that the phenomenon of
"sexuality" in bacteria bears any direct relationship to the process
of sex differentiation in higher organisms, to which we now return.

Aspects of the relationship between chromosomes, growth, and
sex will be further discussed in Chapter 7. Chapter 6 will deal
with some recent advances in our knowledge of sex chromosomes
and sex differentiation in man and other mammals.

# Chapter 6

## Sex Determination in Man and Other Mammals

### I. Introduction

During the first half of the century, ideas regarding the sex chromosomes in man were largely based on data obtained from other organisms. Only a few investigators had actually seen human chromosomes, notably Painter (1924), who described the association of the X and the Y chromosome during metaphase of the first meiotic division of the spermatocyte.

In recent years, however, advances in the field of human chromosomes, and of sex chromosomes, in particular, have come into the forefront of cytogenetic research. The turning point came in 1956, when, following the publication of the chromosome number in man by Tjio and Levan, Ford and Hamerton (1956) examined human spermatocytes using modern techniques and demonstrated the existence of an X and a Y chromosome.

In the years which followed, attention was focused on the somatic chromosomes, and several investigators published arrangements of chromosomes in which either individual pairs or groups of pairs were distinguishable by their relative size and the position of the centromere. Such arrangements came to be known as "karyotypes." The human karyotype was standardized by the Denver Conference (1960) (Fig. 6.1). The sex chromosomes of the female consist of a pair of medium-sized chromosomes with the centromere

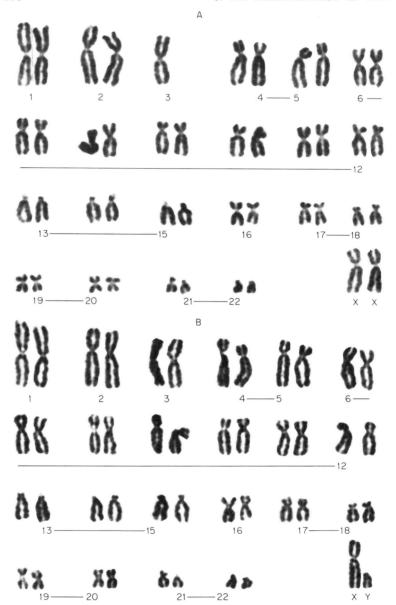

*Fig. 6.1.* Human karyotype, as seen with conventional stains. (A) Female; (B) male (acetic orcein, × 2,200). (Contributed by J. D. A. Delhanty.)

near the middle (usually called "submetacentric"). In the male, there is only one X chromosome and a much smaller Y chromosome, whose centromere is near one end (acrocentric). Results obtained with the aid of the new staining techniques (Chapter 4, Sections XII and XIII) have been incorporated in the standardization agreed on by the Paris Conference (1973).

With ordinary cytological stains, the X chromosome cannot be individually distinguished, since it resembles chromosomes 6–12 (C group) in size and shape. There are, therefore, sixteen such chromosomes in normal females as compared to fifteen in males. The Y chromosome resembles chromosomes 21 and 22 (G group), so that there are five such chromosomes in males and only four in females. The Y chromosome is, however, recognizable in a large proportion of preparations, since the two chromatids of its long arm tend to lie very closely together and to have fuzzy distal parts, while its short arm is very short indeed and lacks satellites.

By using autoradiography, one of the two X chromosomes of females can be identified since it takes up far more tritiated thymidine at a late stage of DNA synthesis than do the other chromosomes (Chapter 4, Section XI).

As a result of the recently developed techniques of fluorescence microscopy and selective Giemsa staining, all the human chromosomes can be individually identified (Figs. 4.4 and 4.6). The centromeric region of the X chromosome stains prominently after treatment. This treatment involves first the denaturing then the renaturing of the DNA and then staining with Giemsa (Arrighi and Hsu, 1971; Schnedl, 1971; Sumner *et al.*, 1971b). On the whole, the same regions stain darkly with Giemsa and fluoresce with quinacrine dyes. The distal (i.e., furthest from the centromere) part of the Y chromosome, which fluoresces particularly strongly (Zech, 1969; Pearson *et al.*, 1970), also stains very darkly with Giemsa. With both techniques, the two X chromosomes of females appear indistinguishable.

The Y chromosome is subject to certain variations. Some of these may be regarded as normal, while others preclude normal development. The Y chromosome may vary in length. Among 517 males examined by Court Brown *et al.* (1965), twelve had Y chromosomes which the authors regarded as outside the normal range of variation. These included large Y chromosomes which were as large as, or larger than, chromosomes 19 and 20, and small Y chromosomes which were not more than one-half the length of chromosomes 21 and 22. Y chromosomes of different lengths are transmitted unchanged from father to son and are not necessarily associated with any detectable abnormality of development.

Jacobs and Ross (1966) reported two patients with female phenotype and streak gonads (see below) who had a structurally abnormal metacentric chromosome, which was interpreted as an isochromosome of the long arm of the Y chromosome (i.e., two long arms and no short arm). There was no evidence of the presence of another cell line, and the authors concluded that the male-determining region of the Y chromosome is located on the short arm. The finding by Lucas and Dewhurst (1972) of a female patient with streak gonads, who had a translocation involving the fluorescing portion of the long arm of the Y chromosome and autosome 4 or 5, would seem to support this view. It must be borne in mind, however, that similar female phenotypes may result in the presence of apparently normal Y chromosomes (see this chapter, Section VII).

## II. Klinefelter's and Turner's Syndromes

Soon after the chromosome constitution of normal males and females was established, the existence of a large number of sex chromosome abnormalities became known.

Patients with Klinefelter's syndrome have male internal and external genitalia, but the testes are small and lack spermatogenesis (Fig. 6.2); enlarged breast development may or may not be present and there is a tendency toward mental retardation (Klinefelter et al., 1942; Overzier, 1963a). The 47, XXY chromosome constitution, first described by Jacobs and Strong (1959), is the most common one in Klinefelter's syndrome. Indeed, with an incidence of 2 cases per 1,000 male births (Hamerton, 1971b), it is the most common sex chromosome abnormality in man. Patients with Klinefelter's syndrome have sex chromatin in buccal mucosa and other cells.

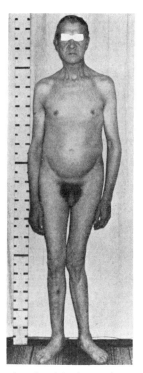

*Fig. 6.2.* Patient with Klinefelter's syndrome. (From C. Overzier, "Die Intersexualität," Fig. 98c, p. 290. Georg Thieme Verlag, Stuttgart, 1961.)

Variant sex chromosome constitutions may occur in essentially the same clinical syndrome. Thus, patients with 48, XXXY chromosomes (Barr *et al.*, 1959) do not differ markedly from 47, XXY subjects though they may be somewhat more severely affected. They have, however, two sex chromatin bodies in a proportion of their cells. Patients with 49, XXXXY chromosomes (Fraccaro and Lindsten, 1960; Barr *et al.*, 1962; Zaleski *et al.*, 1966) have additional abnormalities, including skeletal malformations, underdevelopment of the genitalia, and more severe mental deficiency.

Other chromosome constitutions found occasionally in Klinefelter's syndrome or related conditions include 48, XXYY and 49, XXXYY. Roughly one-quarter of patients are mosaics with two or more cell lines, the chromosome combinations 46, XX/47, XXY and 46, XY/47, XXY being the most common in this group.

In a very small proportion of patients with Klinefelter's syndrome no Y chromosome has been detected at all. These so-called "XX males" will be discussed in Section V of this chapter.

Apart from Klinefelter's syndrome, the best known sex chromosome abnormality is that in Turner's syndrome (Fig. 6.3). In this condition, the three most characteristic abnormalities are rudimentary gonads, short stature, and short neck, with webbing (Turner, 1938; Hauser, 1963a; Lindsten, 1963; Hamerton, 1971b). Instead of ovaries, these patients have only ridges of whitish tissue,

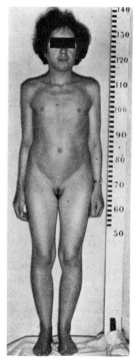

**Fig. 6.3.** Patient with Turner's syndrome. (From G. A. Hauser, "Die Intersexualität," Fig. 105a, p. 310. Georg Thieme Verlag, Stuttgart, 1961.)

which are generally referred to as "streak gonads" (Wilkins and Fleischman, 1944). Another name for the syndrome is "ovarian dysgenesis." The 45, X chromosome constitution, first described by Ford et al. (1959), is the most common one in this condition. Other chromosome constitutions encountered include those in which a second X chromosome is present but structurally abnormal and a variety of mixed chromosome constitutions such as 45, X/46, XX and 45, X/46, XY. No sex chromatin is found, unless a structurally abnormal X chromosome, or a normal 46, XX cell line, is present.

A large proportion of conceptions with a 45, X chromosome constitution die *in utero* and are spontaneously aborted (Carr, 1971b). Singh and Carr (1966) reported that the ovaries at first seemed normal and contained germ cells but that these became abnormally few in number after the third month of pregnancy and had disappeared entirely by, or shortly after, birth.

## III. Females with Multiple X Chromosomes and Males with XYY Chromosomes

Both Klinefelter's and Turner's syndromes had been described as clinical entities before the chromosomal nature of the defect was known. Subsequently, as a result of chromosome studies, certain sex chromosome abnormalities were discovered in patients which do not exhibit any recognizable clinical syndrome.

Jacobs et al. (1959a) described a female patient with some sexual abnormalities who had 47, XXX chromosomes. The same chromosome constitution has been described many times, but the phenotypic effects have been found to vary (Barr et al., 1969). Some of these women suffer from primary or secondary amenorrhea, but others are known to have had children. The majority of these had normal male or female chromosomes but a few XXX or XX/XXX women have been reported to have had children with XXX or XXY chromosomes, some of whom were themselves mosaics (see also Baikie et al., 1972; Geisler et al., 1972). As in Klinefelter's syndrome, there is a definite tendency toward mental retardation in women with three X chromosomes; both conditions are more frequently discovered in hospitals for the mentally retarded than at birth.

Females with 48, XXXX and with 49, XXXXX chromosomes have also been reported (Hamerton, 1971b; Kesaree and Woolley, 1963), although the latter condition, in particular, is very rare. It seems that with increasing numbers of surplus X chromosomes the risk of severe mental retardation and multiple malformations also increases.

The 47, XYY chromosome constitution became well known following the finding by Jacobs *et al.* (1965) of 7 men with this karyotype among a sample of 197 inmates of an institution for subnormal male patients with "dangerous, violent, or aggressive propensities." The XYY patients were also unusually tall. Subsequently, 9 men with this chromosome constitution were found among 315 men in a maximum security hospital (Jacobs *et al.*, 1968). These XYY males had no obvious physical abnormality but their mean height was 181.2 cm, compared with a height of 170.7 cm for males in the same hospital who had only one X chromosome. The intelligence of the XYY men did not seem to differ from that of XY males in the same institution. It is possible, however, that undescended testes and other manifestations of hypogonadism may be more prevalent in XYY than in normal males (Court Brown, 1968). The link between "aggressive" behavior and the possession of an additional Y chromosome became less certain when it was discovered that the incidence of the XYY condition at birth is about 1 in 700 newborn males (Ratcliffe *et al.*, 1970). The fate of the large majority of XYY males is thus far unknown. Nevertheless, there is good evidence that this chromosome constitution results in increased stature. Casey *et al.* (1971) found 32 men with 47, XYY chromosomes among 810 inmates of two hospitals caring for mentally subnormal patients requiring special security measures. The mean height of the XYY males was 12 cm more than that of the males with normal chromosomes.

Spermatogenesis has been studied in a number of XYY males and it has been found that the great majority of primary spermatocytes contained a normal XY bivalent (Evans *et al.*, 1970; Hultén, 1970). It may be assumed, therefore, that the additional Y chromosome has been lost prior to the first meiotic division (Ford, 1970a). Using the fluorescent technique, Diasio and Glass (1970) reported that 5% of the sperm from an XYY man showed two fluorescent bodies, while one body was visible in 70% of the sperm.

Satisfactory evidence of reproductive ability is available in a number of cases. Court Brown (1968) cites three XYY males who had a total of eighteen children. All eight male children, in whom the chromosomes were analyzed, were found to be normal. However, Sandquist and Hellström (1969) have reported an XYY chromosome constitution in a father and son.

## IV. The Male-Determining Function of the Mammalian Y Chromosome

The various sex chromosome abnormalities have shown that the human Y chromosome wields a decisive influence in the process of sex determination. If a Y chromosome is present, the phenotype is male, and if there is no Y chromosome the phenotype is female. By contrast, the number of X chromosomes present does not play nearly as important a role in sex determination, even though if an abnormal number of X chromosomes is present, various developmental abnormalities are likely to result. The presence of too many Y chromosomes may also lead to abnormalities, but probably to a lesser extent.

As will be shown below (this chapter, Section XI), the Y chromosome has been shown to have a positive male-determining function in other mammalian species as well (Ford, 1970b).

The male-determining power of the human Y chromosome contrasts with the function of the Y chromosome in *Drosophila melanogaster,* which does not determine whether a fly develops male or female characteristics but merely whether the sperm is going to be normal (Chapter 1, Section VII). Nevertheless, the human species was not the first in which a male-determining Y chromosome was found, for this mechanism had long been known to operate in a flowering plant, the red campion, *Silene,* which had previously been known under the generic names of *Melandrium* and *Lychnis* (Warmke, 1946; Westergaard, 1958).

## V. XX Males

Notwithstanding the very strong evidence in favor of the male-determining power of the human Y chromosome, there are some apparent exceptions to the rule that a Y chromosome is necessary to initiate the development of the male phenotype. One such exception is provided by males in whom only 46, XX chromosomes are detectable.

Males with two X and no Y chromosomes are rare. The condition was reviewed by de la Chapelle (1972), who included 45 cases. The incidence at birth seems to be less than in 1 in 10,000 males.

The XX males resemble patients with Klinefelter's syndrome in having small testes in which no spermatogenesis takes place. The secondary sexual characteristics are more or less normal, though some have undescended testes or enlarged breast development. It seems that 46, XX males are on an average shorter than 47, XXY males, but they are taller than 46, XX females.

The XX male condition could theoretically be due to a number of causes and three principal mechanisms have been proposed to explain its origin.

One of these postulates a mutation at an autosomal locus affecting sex determination. There is evidence in mice as well as in goats that changes in an autosomal gene (or a small chromosomal rearrangement) may result in masculinization, including testicular development, of chromosomal females. However, an autosomal change of this type would result in a familial incidence of the condition in the sibs and cousins of affected patients, and this is not found in practice. This makes a genetic causation unlikely in the majority of cases.

The second hypothesis assumes that a Y chromosome, or at least part of it, is present but is translocated onto another chromosome. That a portion of the Y chromosome is translocated on an autosome is a theoretical possibility but is unsupported by any type of evidence. The possibility that the entire Y chromosome is translocated in this way is disproved by the lack of a fluorescent Y body in XX male patients (George and Polani, 1970; Fraccaro et al., 1971).

The a priori hypothesis which is more likely is that part of the

Y chromosome is translocated on one of the X chromosomes was proposed by Ferguson-Smith (1966). The most likely time for this to happen is during the prophase of the first meiotic division of the spermatocyte, when the X and the Y chromosome are associated but normally segregate without crossing over. The fertilization of an ovum by a sperm carrying an X chromosome with the male-determining portion of the Y chromosome attached to it would result in an XX zygote with potentialities for male development.

The subsidiary hypothesis that the locus of the $Xg$ blood group on the X chromosome would be likely to be included in an interchange between the X and the Y chromosome involving the male-determining region of the Y chromosome was originally thought to support the interchange hypothesis. More recent data have shown, however, that the proportion of XX males showing the $Xg(a+)$ phenotype is very similar to that of XXY males and significantly different from that found in normal males (Sanger *et al.*, 1971). Furthermore, it could be shown in four families that the XX sons had not received an $Xg$ gene from his father. This suggests that the father had originally contributed a Y chromosome, which initiated male development, but was subsequently lost or has become very rare.

The last finding leads to the third hypothesis, which assumes that a Y chromosome is present in some of the cells of these patients or at least was present at an early stage of their development; subsequently the Y chromosome may become lost or may be present only in a small minority of cells. Forgetting for a moment the possibility of mixed cell lines, an XX chromosome constitution could arise either from a normal XY or from an XXY zygote. Evidence in favor of the latter possibility is presented by four XX males (cited by de la Chapelle, 1972), in whom at most 14 cells with XXY chromosomes were found out of a total of more than 1300 cells examined. With a distribution of this kind one would, of course, expect to find no XXY cells at all in a large proportion of cases.

These findings, therefore, support the evidence from $Xg$ blood groups that a substantial proportion of XX males originate from XXY zygotes. The fact that the Y chromosome may sometimes be lost in these circumstances is not really surprising. Jacobs *et al.* (1963) reported that in blood cultures from normal males, the frequency with which a Y chromosome was lost increased with the age of the donors. In men over 75 years, $6\frac{1}{2}\%$ of cells had less than 46 chromosomes and most of these cells could be accounted for by a missing Y chromosome. It seems likely that this tendency for the Y chromosome to be lost is increased in cells with XXY chromosomes.

## VI. True Hermaphroditism

In clinical terminology, a true hermaphrodite is a person whose gonads contain both ovarian and testicular tissue. Lateral hermaphrodites have a testis on one side and an ovary on the other, bilateral hermaphrodites have ovotestes on both sides, while unilateral hermaphrodites have an ovotestis on one side and a testis or ovary on the other (Overzier, 1963b). Clearly, however, the complete diagnostic data are not always available. The external genitalia tend to be ambiguous (Fig. 6.4). Some hermaphrodites are raised as males, others as females.

Of 108 cases cited by Polani (1970), 59 had 46, XX chromosomes while 21 were 46, XY; 28 cases had more than one cell line, and of these 26 contained a Y chromosome while two did not. Thus, in the majority of hermaphrodite patients, only XX cells have been detected, and once again the question arises of why testicular tissue should develop in the absence of a Y chromosome. As in XX males, the evidence suggests that undetected mosaicism also occurs in at least a proportion of XX hermaphrodites.

Of the patients with mixed cell lines, the most interesting from a scientific point of view are those with 46, XX and 46, XY cells. In some of these there is good evidence that they must have arisen by a process of double fertilization, involving an X- and a Y-bearing sperm and two female haploid nuclei. The first such patient, described by Gartler *et al.* (1962), was a unilateral hermaphrodite, who had two populations of red cells which differed in their *MN* and *Rhesus* blood groups; five additional XX/XY hermaphrodites with two populations of red cells are listed by Benirschke (1973). Mixed red cells, however, cannot always be detected in these her-

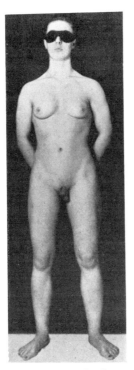

*Fig. 6.4.* Patient with true hermaphroditism. (From C. Overzier, "Die Intersexualität," Fig. 70a, p. 200. Georg Thieme Verlag, Stuttgart, 1961.)

maphrodites. In five XX/XY hermaphrodites enumerated by Fitzgerald *et al.* (1970)—the cases partially overlap with those included by Benirschke (1973)—the total number of XX cells counted was 658, compared with 298 XY cells. In one patient, an excess of XY cells were found, in another patient, the two cell lines were about equal, while in three patients, the XX cells were in the majority. In one of these, a child with unilateral hermaphroditism described by Fitzgerald *et al.* (1970), only five XY cells were found compared with 411 XX cells. Altogether four blood and three skin samples were examined and a few XY cells were found in two skin samples. In another child with lateral hermaphroditism, described by Brøgger and Aagenaes (1965), chromosome analyses from bone marrow, skin, blood, and testis revealed only

XX chromosomes, whereas, in a second testicular biopsy, a proportion of XY cells were found.

In the light of this evidence it seems likely that at least some, if not all, of the hermaphrodites in whom only XX cells have been found also had an XY line, which had either become very rare or been lost completely during the course of development.

Thus, the findings in the rare Klinefelter-type males with apparent XX chromosomes, as well as in hermaphroditism, actually lend further support to the strong evidence in favor of the male-determining power of the mammalian Y chromosome by indicating that even if no Y chromosome can be detected, it is likely to have been present at an earlier stage of development and initiated testicular development in at least a proportion of cases. Whether there is a residual group of patients in whom testicular development took place in the absence of a Y chromosome is still not known for certain. The fluorescent staining technique will undoubtedly be applied more frequently in the future, but if no fluorescent Y bodies are found, no conclusion can, of course, be drawn regarding their absence in the embryo.

## VII. Gonadal Dysgenesis

The presence of a mixture of XX and XY cells does not necessarily give rise to hermaphroditism. Forteza et al. (1963) reported it in a girl with streak gonads and infantile genitalia; no other abnormalities were seen. This condition is often referred to as "pure gonadal dysgenesis" (see below). The patient described by Bain and Scott (1965) was a 24-year-old girl with reasonably well-developed secondary sexual characteristics. In place of gonads she had a streak on the right side and a solid tumor, 13 cm in diameter, on the left, which was classified as a dysgerminoma. Patients with streaks on one side and an ovary or testis on the other have been placed in the category of "mixed gonadal dysgenesis" (Sohval, 1964). The sex chromosome constitution is variable, although 45, X/46, XY is common. The appearance of the external genitalia is also variable, and, as in true hermaphrodites, some patients present as phenotypic males, others as females.

By contrast, patients with pure gonadal dysgenesis, who have streak gonads or may lack gonads altogether, present as females. Two patients with this condition were described by Swyer (1955); the term "pure gonadal dysgenesis" was proposed by Hoffenberg and Jackson (1957). The condition is not an entirely clear-cut entity; for instance, there is a tendency to include patients whose height is 152 cm or above in this syndrome and those below this height in Turner's syndrome. Attempts to correlate the different clinical manifestations of the gonadal dysgenesis with the underlying chromosome constitutions have recently been made by German (1970) and by Polani (1970).

Hamerton (1971b) lists the chromosome constitutions of 40 patients with pure gonadal dysgenesis as follows: 21 with 46, XX; 9 with 46, XY; 2 with 45, X/46, XX; 3 with 45, X/46, XY; and 5 with structural abnormalities of the X chromosome.

The female phenotype of the patients with XY chromosomes may be explained by the failure of the Y chromosome to induce sufficient testicular development. Teter and Boczkowski (1967) report a high incidence of gonadal tumors in these patients; the authors postulate that the tumors originate from embryonic germ cells which persist in the misplaced testicular tissue.

Instances of several cases of pure gonadal dysgenesis occurring in the same sibship have been reported (see Hamerton, 1971b).

It is evident that the different varieties of gonadal dysgenesis as well as true hermaphroditism may be caused by different chromosome constitutions; conversely, the same chromosome constitution may give rise to different pathological entities. The causes underlying these events are so far unknown.

## VIII. Testicular Feminization

By comparison with the conditions just described, the syndrome of testicular feminization forms a well-defined entity, both from the clinical and the genetic point of view. However, there is ample speculation, backed by a good deal of experimental evidence, as to its causation.

The name "syndrome of testicular feminization" was applied by

Morris (1953), who described 2 patients and reviewed an additional 79 cases from the literature. Patients with this condition have a typically female appearance with well-developed breasts, though they lack axillary and pubic hair (Fig. 6.5). They do not menstruate and are sterile (Hauser, 1963b; Polani, 1970). On the average, they seem to be taller that normal women.

In spite of their feminine appearance, these patients have testes, which are usually located in the abdomen, although sometimes

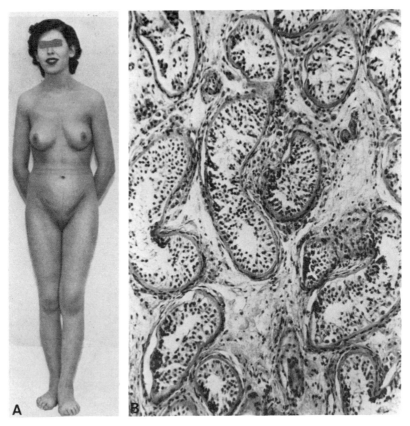

*Fig. 6.5.* Testicular feminization. (A) Patient; (B) histology of testis. [From W. Lenz, *In* "Humangenetik" (P. E. Becker, ed.), Vol. III, Fig. 32, p. 381. Georg Thieme Verlag, Stuttgart, 1968; by permission of H. Nowakowski.]

in the inguinal canal. The testes are rather small and usually no spermatogenesis occurs, but the presence of sperm has been reported in a few cases. There are no vasa deferentia or uterus, and the vagina ends blindly. Jacobs *et al.* (1959b) found the chromosome constitution of four cases to be 46, XY, and this karyotype has since been reported many times.

Just as the gonads in female patients with 46, XY gonadal dysgenesis, the testes in patients with testicular feminization have an increased risk of becoming malignant, so that many medical authorities are in favor of removing them prophylactically. However, if they are removed before puberty the breasts fail to develop. It is thought that the breasts normally develop in this syndrome as a response to estrogen secreted by the testes.

Although it appears that the testes secrete testosterone at concentrations approaching the lower levels of normal males, this hormone fails to effect any masculinization (Polani, 1970). The idea that the underlying fault is a poor or absent response of the target organs to testosterone was expressed by Wilkins (1965), who failed to obtain hair growth in the pubic skin of a patient after injecting testosterone or methyltestosterone.

The condition shows a marked familial aggregation and tends to reoccur not only among sibs but also in the mother's sibship and among maternal cousins. The pattern of inheritance is compatible with a gene location either on the X chromosome or on an autosome; in the latter case, it would be presumed that the abnormal allele acts only in chromosomal males. It has not been possible to make this distinction so far in human data, since patients with testicular feminization do not reproduce and no genetic linkages have been established. However, Lyon and Hawkes (1970) discovered a condition in the mouse which resembles testicular feminization in humans, and it could be shown that the gene responsible in the mouse is linked to other genes on the X chromosome. This finding increases the probability that the gene for testicular feminization in man is also located on the X chromosome.

Ohno (1971) has presented evidence showing that while the proximal tubule kidney cells of normal mice respond to testosterone by producing alcohol dehydrogenase and $\beta$-glucuronidase, mice with testicular feminization fail to produce these enzymes. He postulated that the normal allele of the testicular feminization gene acts as a regulator gene responsible for the development of the secondary sexual characteristics. This theory will be discussed in Chapter 7 (Section XII).

## IX. Tumors Affecting the Secondary Sexual Characteristics

It has already been mentioned that the testes in patients with testicular feminization as well as the gonads of persons with gonadal dysgenesis have an increased risk of becoming malignant. The magnitude of the risk is difficult to evaluate because of general problems of sampling and because patients who have developed tumors are more likely to be reported (Overzier and Hoffmann, 1963). The nature and origin of some of these tumors has been discussed by Hughesdon and Kumarasamy (1970). Tumors of the gonadoblastoma type are extremely rare in normal males and females and are almost confined to gonadal dysgenesis patients having a Y chromosome. These tumors tend to have masculinizing effects, but occasionally the production of estrogen seems to prevail, resulting in enlargement of the breasts.

Very rarely, different types of ovarian tumors may occur in hitherto normal women and lead to amenorrhea, hirsutism, and a deepening of the voice, while certain testicular tumors cause breast development and a feminine hair distribution because of an overproduction of estrogens. Essentially similar feminizing effects may also occur as a result of tumors in the adrenal cortex of males (Overzier and Hoffmann, 1963). On the other hand, tumors of the adrenal cortex may also result in masculinizing effects in both males and females (Bierich, 1963). In the adrenogenital syndrome, the genitalia and the entire physique are masculinized because of an overproduction of androgens by the adrenal glands. Although tumors are frequently present when the disease is manifested in later life, this condition is not primarily a malignant one.

It is clear that abnormal growth processes in the gonads, irrespective of whether the disturbance originates in the gonads themselves or is imposed upon them by an abnormal growth of the adrenal cortex, may have profound effects on the development of the sec-

ondary sexual characteristics, which may occur in a direction contrary to that indicated by the sex chromosome constitution.

## X. Triploidy and Other Autosomal Effects

Triploid karyotypes, with 69 chromosomes in all, are usually
inviable in the human species. Although Carr (1971a) has estimated
that about 1% of all conceptions result in a triploid zygote, the
condition is extremely rare at birth. The large majority of triploid
embryos die within the first 3 months of pregnancy, while the few
which survive birth usually die within the neonatal period.

The reason for this lethality is, at present, unknown and may
seem particularly surprising in view of the fact that triploidy is
a well-known condition in plants, *Drosophila,* and even amphibia.
Mittwoch and Delhanty (1972) have presented evidence that
human triploid fibroblasts may be slower to initiate DNA synthesis
and complete mitosis than diploid ones. If this should prove to
be a general characteristic of triploid cells, it would result in growth
retardation and might explain triploid inviability in man and other
mammals in contrast to lower organisms; clearly, mammals have
a far more stringent developmental schedule than other organisms.

Human triploidy is of special interest in view of the association
of the 69, XXY karyotype with intersexuality (Butler *et al.,* 1969;
Chambon, 1972). The latter author included 7 cases of live births
with this chromosome constitution and no evidence of an additional
diploid cell line (Edwards *et al.,* 1967; Bernard *et al.,* 1969;
Schindler and Mikamo, 1970; Schmickel *et al.,* 1971; Pratt *et al.,*
1971; Leisti *et al.,* 1970; Keutel *et al.,* 1970). All the triploid
children had maldeveloped genitalia; the abnormalities included
undescended testes, hypospadias, ambiguous external genitalia,
and persistent uterus. In a further case, the penis and scrotum were
smaller than normal and the testes, also small, were incompletely
descended (Simpson *et al.,* 1972). Intersexuality has also been described in patients who had a mixture of diploid and triploid cells.
Although in cases with 69, XXY/46, XX chromosomes this could
theoretically be accounted for by the presence of the diploid XX
cell line, it is of interest to note that a child with 69, XXY/46, XY

chromosomes, described by Ferrier *et al.* (1964), had a very small penis and scrotum and two undescended testes.

The occurrence of intersexuality in an 69, XXY karyotype is of particular interest in view of the fact that the 47, XXY karyotype results in unambiguous male development. It seems, therefore, that the additional number of autosomes prevents the single Y chromosome from imposing full masculinity on the developing phenotype; this, in turn, might mean that the growth rate of the gonadal rudiment might be insufficient to ensure the development of adequately functional testes (see Chapter 5, Section VII).

It will be remembered that in *Drosophila melanogaster,* triploid flies with XXY chromosomes are also intersexual and that this was explained by postulating the existence of male-determining genes on the autosomes (Chapter 1, Section VII). It is possible that in *Drosophila,* the differentiation of the female phenotype may require a faster rate of growth than that of the male, in contrast to the situation in mammals.

Ambiguous genitalia have been reported in a human infant with 46, XY chromosomes, in whom two chromosomes of the C group showed a reciprocal translocation (German and Simpson, 1971). In autosomal trisomies with XY chromosomes, the phenotype is usually unambiguously male; nevertheless, Taylor (1968) reported the incidence of undescended testes in 4 out of 4 males with Edward's syndrome (trisomy 18) and 13 out of 14 males with Patau's syndrome (trisomy 13).

## XI. Sex Chromosomes in Other Mammals

In the large majority of mammalian species, the sex chromosome mechanism is the same as in man, i.e., XY in males and XX in females (see Hsu and Benirschke, 1967). These include the commonly used laboratory mammals such as the mouse (*Mus musculus*), the rat (*Rattus norvegicus*), and the rabbit (*Oryctolagus cuniculus*) and domestic animals such as the horse (*Equus caballus*), cattle (*Bos taurus*), sheep (*Ovis aries*), goats (*Capra hircus*), and the pig (*Sus scrofa*), and the cat (*Felis domesticus*) and the dog (*Canis familiaris*).

In some species, certain variations of the sex chromosomes occur, usually in size or in number (Fredga, 1970). The short-tailed vole (*Microtus agrestis*) has giant sex chromosomes, the X and the Y chromosomes being the two largest in the karyotype. The common shrew (*Sorex araneus*) has two Y chromosomes, the sex chromosomes being XX in females and $XY_1Y_2$ in males. This mechanism is thought to have arisen through translocation of the X chromosome onto an autosome so that the sex chromosome constitution could be designated as $A^XA^X/A^XYA$. By contrast, species with multiple X chromosomes, in which the chromosome mechanism is $X_1X_1X_2X_2/X_1X_2Y$, are thought to have originated through translocation of the Y chromosome onto an autosome; this could be designated as $XXAA/XAA^Y$. The sex chromosomes of the mongoose (*Herpestes auropunctatus*) are probably of this type; a trivalent is formed in meiosis of the spermatocyte.

There is every indication that the Y chromosome is male-determining throughout the order of mammals. The existence of mice with a single X and no Y chromosome was first described by Russell *et al.* (1959); these mice are female and fertile. Although apparently less phenotypically abnormal than their human counterpart, XO mice produce only about one-third of the expected proportion of XO daughters (Welshons and Russell, 1959; Cattanach, 1962; Morris, 1968). It is thought that this is at least partly due to an increased mortality of XO embryos, although nonrandom segregation of the X chromosome in the oocyte might be an additional factor.

A mouse with XXY chromosomes was described by Russell and Chu (1961) and two others by Cattanach (1961). All three animals were sterile males with small testes. Similarly, sterile males have been described in a number of other species, including the dog (Clough *et al.*, 1970) and the pig (Breeuwsma, 1968). This chromosome constitution also occurs in male tortoise-shell cats, although often in conjunction with other cell lines. The situation in the tortoise-shell tomcat, which had been puzzling geneticists for many years, was recently reviewed by Jones (1969). The tortoise-shell phenotype is produced by combining the gene for yellow ($y$) with its nonyellow ($y^+$) allele. Since the gene is borne on the X chromosome, males are normally either yellow (ginger or orange) or nonyellow, usually black or tabby. Tortoise-shell male cats are sterile. Thuline and Norby (1961) found 2 such cats with 39 chromosomes, presumed to be XXY. The chromosome constitutions of an additional 17 tortoise-shell male cats are given by Benirschke (1973). Only one of these had an apparently single XXY cell line; five were diploid/triploid, probable chimaeras (see below), of whom four were 38, XX/57, XXY and five were 38, XY/57, XXY; four cats were 38, XX/38, XY (one of these was a true hermaphrodite); four others were 38, XY/39, XXY; one cat was 38, XY/38, XY, with two different cell populations (this cat was probably fertile); the remaining two had more than two different cell lines, being 38, XX/38, XY/39, XXY/40, XXYY and 38, XY/39, XXY/40, XXYY, respectively. Of the four 38, XX/38, XY cats, three had testes and only one was a true hermaphrodite; this cat was not tortoise-shell but had two eyes of different colors.

Intersexes with mixed XX and XY cells have also been reported in other domestic animals, including cattle and pigs (McFeely *et al.*, 1967).

## XII. Sex Chromosome Chimaeras

The origin of mixed cell lines with different chromosome constitutions are varied. A basic distinction is whether the organisms arose from a single zygote cell line or by fusion of two or more zygote cell lines. It is now becoming accepted practice to use the term "mosaic" for the first group and "chimaera" for the second group (Ford, 1969; Benirschke, 1973). In practice, however, the origin of mixed cell lines is not always certain. For instance, in human hermaphrodites with mixed cell lines containing XX and XY chromosomes, as well as a mixed red cell population containing two different phenotypes of blood group antigens, the existence of more than one product of fertilization may be taken as proved beyond reasonable doubt; but in the absence of genetically different red cell populations such a conclusion would be less certain, since a mixture of XX and XY cells could theoretically arise by mitotic nondisjunction from, for example, XY or XXY cells.

In the mouse, chimaeras have been produced artificially by fusing blastocysts *in vitro* (Tarkowski, 1961; Mintz, 1964). Two 8-cell embryos can be combined in this way to form a single organism, which will usually contain both genetically distinct cell types. Since the sex chromosome constitution of the constituent blastocysts is unknown at the time of fusing them, one-half of the resulting chimaeric mice will be composed of XX and XY chromosome, while one-quarter each will be entirely XX or XY, respectively. Those latter mice would be expected to develop as females or as males, while the group with mixed XX and XY cell lines might theoretically develop as hermaphrodites. Hermaphrodite individuals do, in fact, occur among chimaeric mice, but their proportion is far less than 50% of the total (Mystkowska and Tarkowski, 1968; McLaren and Bowman, 1969). It seems that the majority of mice with mixed sex chromosome constitutions develop into apparently normal, fertile males and only a small minority become hermaphrodites. It should be noted that the proportion of XX and XY cells originally present in the fused embryos is not necessarily 1:1, since a proportion of the original cells of the blastocysts become part of the extraembryonic membranes. The sperm formed in XX/XY mice always seems to be derived from the XY cells (Tarkowski, 1970). The question of whether in chimaeric animals of diverse species, germ cells whose sex chromosomes are at variance with their phenotypic differentiation has been actively debated (Benirschke, 1973). It now seems that this is unlikely in mammals, at least as far as sperm formation is concerned. The reason for this is not yet known. It may be surmised, however, that the rather strict requirements in the timing of mitosis and meiosis in mammalian spermatogenesis are incompatible with any but an XY chromosome constitution.

Recent results by McLaren *et al.* (1972) suggest that oocytes also may be unable to continue their development in a testicular environment. The authors examined chimaeric mouse fetuses aged between $15\frac{1}{2}$ and $18\frac{1}{2}$ days and found germ cells in meiotic prophase in the gonads of all female fetuses. Among the males, germ cells in meiotic prophase were seen in 6 out of 23 fetuses, and on day $18\frac{1}{2}$ the meiotic cell were degenerating. It was concluded that the meiotic germ cells were probably XX and that they rarely survive when surrounded by testicular tissue.

In animals with multiple births, hemopoietic cells may be exchanged through vascular anastomoses between different fetuses, thus giving rise to "blood chimaeras" (Benirschke, 1973). This happens regularly in cattle twins and in marmosets (*Callithrix jacchus*), in which twins are regularly born, but occurs only rarely in human twins. Another difference is apparent in the sexual development of heterosexual chimaeras. This is normal in humans and marmosets, as well as in horses, whereas in cattle twins of unlike sex, the female is rendered intersexual and becomes a so-called "freemartin." Because of the historical and current interest of this condition in relation to sex differentiation, the freemartin will be discussed in somewhat more detail.

## XIII. The Freemartin

The freemartin is a sterile twin; this condition is best known in cattle, though the condition also occurs in sheep (Alexander and Williams, 1964), in goats (Ilbery and Williams, 1967), and in pigs (Hughes, 1929). It was shown more than 50 years ago (Lillie, 1917; Tandler and Keller, 1911) that freemartins occur only if the twins are of opposite sex and if vascular anastomoses are present between the twins. Under these conditions the female calf becomes masculinized, and it was postulated that this is due to action of male sex hormone which is secreted by the male twin and enters the female twin through the common circulation.

The external genitalia of the freemartin resemble those of a female but the Mullerian duct derivatives are absent or greatly reduced, while the ovaries are also reduced and nonfunctional and, in addition, may contain varying proportions of seminiferous tubules (Wells, 1962). The formation of ovotestes presents a problem, as the administration of exogenous androgens has not been found to affect the differentiation of the gonads. High doses of androgens injected into cows before the time of gonadal differentiation of the fetus resulted in female calves being masculinized but having normal ovaries. These calves were not freemartins (Jost et al., 1963). In developing freemartins, on the other hand, the ovaries were at first apparently normal and their development became inhibited only after the testes were well differentiated. Evidence for the production of androgens by the gonads of a freemartin was provided by Short *et al.* (1969).

Following the claim by Ohno *et al.* (1962) that in newborn twin calves of opposite sex, germ cells with XX chromosomes are present in the testes of the male twin, there has been speculation whether the freemartin effect may be due to presence of XY germ cells in the gonads of the female twin. However, apart from lack of evidence for this, such an interpretation is fraught with a number of difficulties. The migration of germ cells via a vascular route has neither been established in mammals (Tarkowski, 1970), nor would the sex chromosome constitution of germ cells be expected to affect the differentiation of the gonads; and even if heterosexual germ cells were implicated, it is not clear why they should result in the freemartin effect in cattle and sheep but not in humans and marmosets (Benirschke, 1973).

The freemartin provided the classical natural experiment on which the theory of a hormonal causation of sex differentiation was based. While the theory has withstood the test of time, the causation of the freemartin condition itself remains an enigma.

## XIV. Autosomal Sex Reversal

In the goat, *Capra hircus,* a dominant autosomal gene, polled (*P*)—which might, however, be a small chromosomal rearrangement—has an effect on the growth of the horns. When present in the homozygous state (*PP*), this factor causes the development of testes in chromosomal females, which are also polled, i.e., hornless. The testes do not contain any germ cells at birth, but germ cells are originally present and degenerate during fetal life (Hamerton *et al.*, 1969b, Hamerton, 1971b). Some homozygous males are also sterile as a result of epididymal occlusion.

Another autosomally located sex-reversing factor has recently been reported in the mouse (Cattanach *et al.*, 1971). The *Sxr* factor behaves as a dominant and, when present in single dose, causes

XX animals to develop as males with small testes. These contain male-type germ cells during fetal and early postnatal stages, but they are progressively lost, with none left at 10 days of age.

When the Sxr factor was carried by XO mice, which would normally develop into fertile females, they were transformed into males with near normal-sized testes; in some cases sperm was actually formed. None of the spermatozoa were motile however.

We see, therefore, that both in goats and in mice an autosomal gene—or small chromosomal rearrangement—can bring about testicular differentiation in animals which lack a Y chromosome. It is possible that the autosomal factors may bring about an increase in growth in the gonadal rudiment, and it is of interest to note that, in the mouse, testicular differentiation is more complete in animals having only one, rather than two, X chromosomes.

It was shown by Mittwoch and Buehr (1973) that the volumes of XX, Sxr testes in embryos aged 15 and 16 days were very much larger than those of ovaries and almost, but not quite, as large as those of XY testes. This supports the hypothesis of a causal relationship between a fast gonadal growth rate and testicular differentiation. It further suggests the possibility that small testes lacking spermatogenesis in the adult may be due to a suboptimal growth rate in the embryo.

It may also be noted that XO mice carrying the Sxr factor resemble *Drosophila melanogaster* with XO chromosomes in being phenotypical males with nonmotile sperm. This suggests that in mammals as well as in flies, the timing of the different stages of spermiogenesis is so critical that the normal end result can be achieved only in the presence of a Y chromosome, even if testicular differentiation and the development of male secondary sexual characteristics can be achieved in its absence.

# Chapter 7

## Genes, Chromosomes, Growth, and Sex

### I. Introduction

Any attempt at summarizing our present-day knowledge of the genetics of sex differentiation involves the coordination of pieces of information obtained from a variety of organisms. Our own species has come into the foreground in recent years by revealing an ever growing mass of data regarding the existence of sex chromosome variations and their effect on the phenotype. Following closely in the wake of advances in human cytogenetics, there has been a great increase in our knowledge of the chromosomes of other mammals, and more and more sex chromosome abnormalities are becoming known. By comparison, our knowledge of the sex chromosomes of birds is still meager, for although the sex chromosome constitution of a number of species has now been established, we do not yet know the effect of a numerical aberration of the sex chromosomes on the sexual development of birds. It is to be hoped that this lack of knowledge will be remedied in the near future, so that we may gain a better understanding of the mechanism of female heterogamety in vertebrates.

By contrast, in *Drosophila*, numerical aberrations of the sex chromosomes, and their effects on the phenotype, have been known for half a century, and the work of Bridges is still an important foundation upon which new ideas on the genetics of sex differentiation are being built. Although the embryology of sexual development is less well known in *Drosophila* than it is in vertebrates, *Drosophila* has the advantage of being extremely well-known genetically. A few of these "mutants" are known to be due to structual rearrangements of the chromosomes. For instance, *Bar,* which reduces the size of the compound eye, is caused by a small duplication on the X chromosome, while *Notch,* which acts as a lethal in males and affects the shape of the wings in heterozygous females, is due to a short deficiency in the X chromosome. The fact that the structural nature of these apparent mutations could be detected is due to the presence of giant chromosomes in the salivary glands of *Drosophila* larvae, which show a great deal more detail than ordinary chromosomes. Chromosomal changes of this order of magnitude would be out of reach of a cytogenetic interpretation in man or the mouse.

However, even in *Drosophila* there are a large number of so-called gene mutations which do not show any abnormalities in the salivary gland chromosomes and whose basic nature is therefore unknown. This applies particularly to those affecting the shape and size of the body and its appendages. Since, apart from color changes, differences in shape and size are the most obvious to detect on visual inspection, it is hardly surprising that a large proportion of the mutants in *Drosophila* which have been described fall into this class. It seems hardly necessary to emphasize that before any real progress in the analysis of gene action can be made, it will be necessary to separate this medley of mutants at least to the extent of distinguishing between base substitutions in DNA strands and chromosomal variations of an entirely different order of magnitude. With this purpose in mind, some striking and well-investigated mutations, defining this concept in its wider sense, which influence the external morphology of *Drosophila,* will be briefly discussed.

## II. Homoeotic Mutants in *Drosophila*

The term "homoeosis" was introduced by Bateson (1894) to denote the change of an organ in a segmental series into another, homologous organ, belonging to a different part of the segmental series. A classical example is the regeneration of an antenna after the removal of the eye stalk in decapods.

In *Drosophila*, a number of mutants are known which change the normal appendages of the flies into homoeotic structures. In *bithorax* (Bridges and Morgan, 1923; Lewis, 1964), the flies appear to have two thoraxes, each containing the usual bristles and a pair of wings. The condition is due to the anterior part of the metathorax, which normally carries the halteres or balances (highly modified derivatives of the posterior part of hind wings of other insects), becoming modified into mesothorax, which carries a pair of wings. In *aristapedia* (Balkashina, 1929), the antenna is replaced by the tarsus of a leg (Fig. 7.1); in *tetraptera* (Astauroff, 1929), the halteres are replaced by wings; in *tetraltera* (Goldschmidt, 1940; Villee, 1942), the forewings are replaced by halteres. In *proboscipedia* (Bridges and Dobzhansky, 1932), the oral lobes of the proboscis are changed into antenna- or tarsus-like appendages. In *podoptera* (Goldschmidt *et al.*, 1951), the wings are transformed into leglike structures.

Changes such as those of the mouth parts, or of wings into halteres, are regarded as important differences in the classification of insects, and early geneticists derived some measure of satisfaction from the thought that mutations of apparently single genes could have such far-reaching effects. However, homoeotic mutants exhibit several characteristics which do not seem to fit into a simple scheme.

Although most homoeotic mutations show a simple recessive type of transmission, the extent to which the character is expressed is extremely variable. The expression typically varies from a normal, unchanged organ through various intermediate degrees to the apparently complete substitution of a homologous organ (Fig. 7.1). The proportion of homozygous individuals which fail to express the character is affected to a considerable extent by environmental factors, such as temperature and food supply. The penetrance of *tetraltera* was found to be less than 1% at 29°C, 8% at 25°C, and 15% at 20°C, while at 14°C an average of 36.7% of flies showed the *tetraltera* effect (Villee, 1942). In *tetraptera*, the opposite relationship with temperature was found: at 17°C the penetrance was 0–1%, while 35% of flies showed this effect at 25°C (Astauroff, 1930).

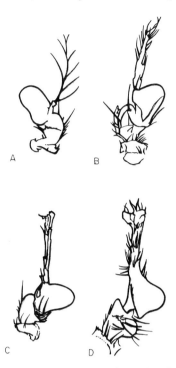

*Fig. 7.1.* Series of *aristapedia* in *Drosophila virilis,* illustrating the different manifestations of the character. (A) Normal antenna and arista; (B) and (C) increasingly leglike modifications of arista; (D) extreme modification of arista, showing strikingly leglike modification. (From Bodenstein and Abdel-Malek, 1949.)

In the case of *tetraltera,* there appears to be no correlation in the penetrance of the character between parent and offspring. The proportion of affected offspring was found to be the same when both parents showed the character, when only one parent showed it, or when both parents appeared normal. Whether or not the character manifests itself in any given fly which is homozygous for the *tetraltera* locus appears to depend on a quantitative variable of environmental origin, with threshold effect (see Chapter 2, Section III). The total time of development of phenotypically normal *tetraltera* flies is 1 day longer than that of normal flies, while flies showing the *tetraltera* character require 2 to 3 days more than normal flies to complete their development. Accordingly, an agency which prolongs development during a critical period early in development will increase the penetrance of *tetraltera* (Villee, 1945).

Another peculiarity which is characteristic of homoeotic as well as some morphological mutants is the lack of symmetry in the manifestation of the character. The correlation in the degree of expression between left and right sides is rather low. Individuals may appear completely normal on one side and show the mutant character on the other.

Yet another striking feature about homoeotic mutants is that five out of the six mentioned have been mapped within a short region of the right arm of the third chromosome. In the case of *podoptera*, no single chromosomal locus could be mapped, as it became apparent that the effect could be produced by various heterochromatic regions in different chromosomes. The Y chromosome was particularly effective in all strains; in some strains, females would show the *podoptera* character only if they carry a Y chromosome (Goldschmidt *et al.*, 1951). This suggests that *podoptera* may be caused by small changes in the heterochromatic regions of chromosomes. In one strain, the *podoptera* factors produced a high incidence of gynandromorphism and chromosome elimination, and Goldschmidt (1949) suggested that changes in heterochromatin might have an effect on chromosome multiplication as well as on chromosome elimination.

It is possible that the basis of other homoeotic mutants may also be in various rearrangements in chromosome 3, which are too small to be seen in the salivary gland chromosomes. The developmental evidence would seem to favor such a view. In mature larvae of *Drosophila melanogaster*, most of the prospective organs of the adult are already laid down in the form of imaginal buds (or discs) of definite size, shape, and structure. A detailed study of the imaginal discs of the eyes, wings, and halteres in normal and mutant flies was carried out by Chen (1929). In the *Bar* mutant, which is due to a visible duplication in the X chromosome (Lindsley and Grell, 1968), the number of eye facets is reduced to between 70 and 80 as compared with about 750 in normal flies. In mature larvae, when the optic buds have reached their maximum size, those of *Bar* larvae were about 45% smaller than the optic buds of normal larvae. The *Bar* eye is due not to any defect in the facets but to a diminution in size of the optic tract.

By contrast, the mutant *bithorax* was found to be associated with an increase in size of the thorax and of the haltere buds. At the end of the larval period, the haltere buds of larvae from *bithorax* stock were more than 50% larger than those of the wild-type larvae.

## III. Phenocopies

We have seen that the expression of homoeotic mutants is sensitive to environmental conditions; as a corollary, certain environmental agents may mimic the development of the phenotype, characteristically seen in those mutants, in individuals which do not carry the mutation.

Bodenstein and Abdel-Malek (1949) succeeded in inducing phenocopies of *aristapedia* in a stock of *Drosophila virilis* by treating larvae with a nitrogen mustard, methylbis($\beta$-chlorethyl)amine hydrochloride. The substance was found to be active only during a definite age range of the larvae. When larvae of 70–80 hours were treated with a 2% solution of the mustard, up to 50% of the flies showed the *aristapedia* effect. As in the case of mutant flies, the expression of the abnormal character was variable; in extreme cases a strikingly leglike structure was formed (Fig. 7.1).

The idea of phenocopies—nonhereditary phenotypes of environmental origin, which mimic the action of mutations—arose from studies of heat shock in *Drosophila* (Goldschmidt, 1935). Subsequently similar effects were obtained as a result of chemical treatment (Goldschmidt and Piternick, 1957a,b). In chick embryos also, a variety of chemical agents were found to be capable of producing phenocopies. For instance, Landauer (1960) reported that if relatively low doses of nicotine were injected into chick embryos, they developed a shortening and twisting of the neck due to incomplete fusion of the cervical vertebrae; the same type of abnormality may also be caused by a recessive lethal mutation "crooked-neck dwarf." Of particular interest is the finding that embryos which were heterozygous for the mutation and phenotypically indistinguishable from normal embryos nevertheless gave a markedly higher yield of the malformation when treated with nicotine. As another example, treatment with the nicotinamide analog, 6-aminonicotinamide induces the development of micromelia and a beak defect known as parrot beak in a large proportion of embryos. Again, the incidence of these malformations was increased when the mothers of the embryos were heterozygous for recessive micromelia mutations (Landauer, 1965), suggesting a relationship between the effect of the gene and of the teratogen. That the common effect may be a reduction in the available cellular energy is suggested by the findings of Landauer and Sopher (1970) that various sources of cellular energy may reverse the effects of teratogens. The additions of succinate and of ascorbate were found to have a strong effect in reducing the incidence of malformations in embryos treated with 6-aminonicotinamide, with 3-acetylpyridine, and with sulfanilamide; glycerophosphate reduced the teratogenic effect of the first two but had no effect in the presence of sulfanilamide. Since succinate, glycerophosphate, and ascorbate are all high-energy intermediate compounds in the process of energy-dependent reversal of electron transport in the respiratory chain, which will function also if supplied from an outside source, Landauer and Sopher conclude that the teratogens in question normally exert their effect by interfering with energy production in the mitochondria or other cytoplasmic organelles. Furthermore, the malformations caused by antimetabolites and other teratogens appear to originate in disturbances of growth rather than of differentiation.

Turning to malformations in mammals, Saxén and Rapela (1969) have presented a scheme of how an inhibition of growth occurring at slightly different stages of embryonic development would result in three different types of interatrial septal defects of the heart. The authors point out that these malformations are common in children whose mothers had rubella in pregnancy. Since there is evidence that the rubella virus may inhibit cell multiplication, the observed malformations of the septa could be a direct effect of the virus. Evidence has also been presented that the risk of congenital heart disease is increased in children of mothers who have been smoking during pregnancy (Fedrick et al., 1971); it has also been shown that women who smoke during pregnancy have children of lower birth weight and somewhat lower intelligence than nonsmoking mothers (Butler and Alberman, 1969). Although the causes of these conditions are likely to be complex (Yerushalmi, 1972), it seems likely that at least some of this is a direct effect of smoking. In this connection it is of interest that congenital heart disease is a common malformation in mongolism (Penrose and Smith, 1966), as well as in trisomy 13 and in trisomy 18 (Taylor, 1968); all three trisomies are associated with low birth weight. Evidence that the addition of chromosomes to the normal karyotype may slow down cell proliferation will be presented in Section VIII (this chapter). It seems, therefore, that either cigarette smoking by the mother or an abnormal chromosome constitution may result in growth retardation, which, in turn, may give rise to congenital heart disease.

A similar interaction between genetic and environmental factors is also likely to occur in anencephaly and spina bifida. These severe developmental defects are partly determined by heredity but not in any straightforward Mendelian manner (Stern, 1960) (see Chapter 2, Section III). Recent evidence presented by Renwick (1972) suggests that the incidence of both conditions is increased by a substance present in potato tubers which have been infected by the blight fungus, *Phytophthora infestans*.

## IV. Genetic Assimilation

The phenomenon of genetic assimilation has been defined by Waddington (1961) as "the conversion of an acquired character into an inherited one; or better, as a shift (toward a greater importance of heredity) in the degree to which the character is acquired or inherited."

Experimental evidence that genetic assimilation does, in fact, occur, was published by Waddington in 1953. Based on the premise that different strains of *Drosophila melanogaster* differ in their capacity for phenocopy formation, a strain of pupae aged 17–24 hours was chosen which ordinarily show a normal phenotype but which, when exposed to temperature shock for 4 hours at 40°C, produced some flies in which the crossveins of the wings are absent or broken. These crossveinless flies were selected as parents, and it was found in subsequent generations that the proportion of crossveinless offspring gradually increased, especially if the temperature shock was given between 21 and 23 hours. After 12 generations of this treatment, over 90% of the flies showed the abnormal crossveins.

In untreated flies of treated parents, crossveinless individuals did not appear until the fourteenth generation, and then only in a few isolated cases. By the fourteenth generation, the proportion of crossveinless flies which had received no heat shock had risen to between 1 and 2%. Pair matings were set up from these, and, by further selection without heat shock, a frequency of 100% crossveinless individuals could be obtained at 18°C, although this percentage was somewhat lower at 25°C. Evidently, therefore, a character which originally developed as a response to an environmental agent, i.e., heat shock, had been converted into an inherited character following selection over a number of generations.

After its assimilation into the genetic material, the production of crossveinless wings could be shown to be affected by many chromosomal regions and was said to be "polygenically" controlled. The X chromosome of selected flies was particularly effective. The X chromosome carries a well known mutant, which causes the development of crossveinless wings when present in a homozygous or hemizygous condition. It is of particular interest that roughly three-quarters of female flies that were heterozygous for both the *crossveinless* allele and the X chromosome which had been selected for producing the crossveinless condition, did, in fact, show the *crossveinless* phenotype. Clearly, the selected X chromosome reinforced the *crossveinless* gene, even though the two did not behave as simple alleles. This suggests a common effect of the gene and the selected chromosome not carrying the gene, similar to that found by Landauer (1965) for the effect of 6-aminonicotinamide and genetic micromelia. Remembering that in the latter case it was suggested that the underlying disturbance was one of growth, it is possible, therefore, that the crossveinless condition may also be due to an altered rate of growth and that the *crossveinless* "allele" may really be a small chromosomal rearrangement, which affects growth. Regarding the process of genetic assimilation,

we may surmise that chromosomal changes with similar effects
were present in the population at the beginning of the experiment
and that both their effects as well as the production of new ones
may have been increased by the heat treatment.

The possible relationship between the genetic assimilation of
acquired characters and the evolution of sex chromosomes will
be discussed in Section XII (this chapter).

## V. Transdetermination in *Drosophila*

Experiments on the differentiation of embryonic cells in *Drosophila,* carried out by Hadorn (1968) and collaborators (Wildermuth,
1970) point to yet another aspect of the interrelationship of the
genetic material and the environment in determining the development of the phenotype.

In the larval stages of higher insects, including flies, groups of
cells are set aside which are destined to give rise to the organs
of the adult, or imago. These cells are known as imaginal discs.
Each disc contains a few thousand cells which remain in an undifferentiated state throughout the larval period. When the larvae
are ready to pupate, their organs disintegrate while the cells of the
embryonic discs develop into the organs of the imago.

Although the cells of the imaginal discs appear undifferentiated,
they are nevertheless predetermined to form particular organs. For
instance, different cells of the leg disc will develop into claws,
tarsal parts, tibia, femur, trochanter, coxa, and adjoining parts of
the thorax, respectively. The positions of the various imaginal discs
in the larva in relation to the organs which they are destined to
form in the adult is illustrated in Fig. 7.2.

If imaginal discs are dissected out and transplanted into another
larva, they will differentiate during the metamorphosis of the host
and form the organs for which they had been originally determined; this differentiation is independent of the location of the
transplanted disc in the body of the host. If, however, the imaginal
discs are transplanted into adult flies, they continue to divide without any sign of differentiating into organs. By transplanting these
proliferating cells every few weeks into new adults, cells of the

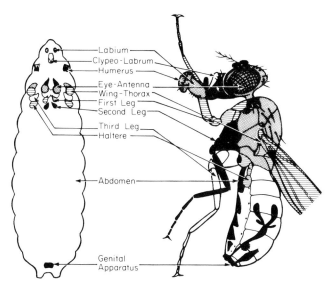

*Fig. 7.2.* Diagram showing topographical situation of imaginal discs in the larva of *Drosophila* and their derivations in the adult fly. (From Wildermuth, 1970.)

imaginal disc have been maintained in an actively growing state for years and appear to divide indefinitely. However, when these cells are subsequently reintroduced into a larva, they will, during metamorphosis, differentiate into adult organs.

The fate of these embryonic disc cells, which have been cultured in this way for a long time, is of particular interest. For a number of transfer generations, depending on the type of imaginal disc, the cells, when introduced into a larva, will differentiate into the organs for which they were originally determined, but after a time, some of the cells will suddenly develop into a different organ. This change has been called "transdetermination."

In a typical experiment, cells of the male genital disc are differentiated exclusively into structures for which it had been determined, i.e., sex organs and parts of the gut and abdomen, during seven transfer generations. In the eighth transfer generation, some of the cells developed into head and leg structures. In the thirteenth transfer generation, some of these cells developed into wings, and in the nineteenth generation into thorax cells. The sequence of transdetermination proceeds in an orderly fashion, as illustrated in Fig. 7.3.

Transdetermination occurs only in cells which are dividing.

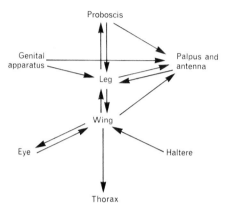

**Fig. 7.3.** Sequence of transformation in *Drosophila*. (From Wildermuth, 1970.)

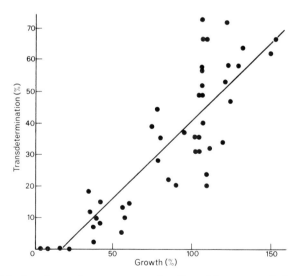

**Fig. 7.4.** Correlation between rate of cell proliferation and frequency of transdetermination in cultured labial discs (From Wildermuth, 1970.)

Moreover, a direct correlation has been observed between the rate of cell proliferation, as estimated either from volume measurements or from counts of bristles in metamorphosed organs, and the rate at which transdetermination occurs (Fig. 7.4).

The fact that the same rudiment may give rise to different organs in accordance with the rate of cell proliferation may be of special relevance to the problem of gonadal differentiation.

## VI. Mitotic Rates and Differentiation

That tissue differentiation may be dependent on preceding mitotic divisions is suggested by several lines of evidence (Tsaney and Sendov, 1971; Bellairs, 1971).

The mature skeletal muscle fiber is a multinucleate unit containing large numbers of cross-striated myofibrils. It is formed by the fusion of large numbers of myogenic cells, each containing a single nucleus. Holtzer et al. (1969) found that in cell cultures fusion of myogenic cells would take place only in the $G_1$ phase, following mitosis. If mitosis was blocked by colcemid or if the cells were treated with 5-fluorodeoxyuridine (FUDR) which blocks DNA synthesis, no cell fusion occurred. However, if following FUDR treatment, the cells were transferred to normal medium, they would form myotubes, provided that they were allowed to undergo one or more cell divisions. The authors suggested that fusion may be dependent on the state of the cell surface and that the proper cell surface may be formed only during a particular $G_1$ following a critical "quantal" mitosis.

The production of zymogen proteins by pancreatic cells is another example of a possible relationship between cell and differentiation and cell division. The differentiation of pancreatic cells in vitro is prevented by FUDR and other inhibitors of DNA synthesis, although after a certain stage of differentiation has been reached, these compounds have no effect on further development (Rutter et al., 1968; Wessels, 1968).

Although the evidence regarding a causal relationship between mitotic divisions and differentiation is not yet conclusive, the results so far are certainly suggestive.

## VII. Chromosomal Volumes and Mitotic Rates

While the causes underlying differential mitotic rates remain shrouded in obscurity, some data on differences in chromosomal volumes within cells of the same organism may turn out to be relevant.

Witschi (1935) found that in the goose barnacle, *Lepas anatifera,* a hermaphrodite crustacean, the chromosomes appear spherical; during the first meiotic division of the oocyte their diameter is nearly twice that of the chromosomes in the corresponding state of the spermatocytes.

Alfert *et al.* (1955) examined nuclei of different sizes in the thyroid glands by means of microspectrophotometry. They found the amounts of DNA and histone to be virtually constant but the total protein content was found to vary in direct relation to the nuclear volume. Moreover, nuclear volumes could be affected by hormone treatment. Hypophysectomized animals had smaller nuclei, while intact animals treated with prophylthiouracil had larger nuclei than untreated controls.

As early as 1937, Pierce reported that treatment of the root tips of *Viola conspersa,* e.g., with varying amounts of phosphates, affected the sizes of the chromosomes. These results were confirmed by Bennett and Rees (1967, 1969) in rye, *Secale cereale.* The volumes of metaphase chromosomes of seedlings kept in "high normal" as well as in the absence of phosphate were found to vary in direct relation to the amount of phosphate. In addition to different treatments, changes in chromosome size were also observed in seedlings of different ages. The different chromosome sizes could be correlated with differences in dry mass, but not with DNA content, which was constant throughout.

In root tips and shoot apices of *Vicia faba,* Bennett (1972) found large differences in chromosomal volumes which were correlated with the ages of the plants and the location of the cells. Again, the chromosomal volumes were positively correlated with the amounts of total nuclear proteins as well as of histones and of RNA, while most of this variation was not related to DNA content.

An important point which emerged was that chromosomal volume showed a positive correlation with mitotic index. Similarly, Lyndon (1968) reported that in *Pisum* in which the cells of the shoot meristem divide more slowly than those of the root meristem, prophase nuclei of the shoot are only about one-half the size of those of roots, with a corresponding decrease in dry weight. The nuclei of the shoot, therefore, synthesize less nuclear protein and RNA than do the nuclei in the root. It thus appears as if the DNA of the chromosomes forms complexes with protein and with RNA, which vary in relation to the mitotic rate. It would appear, moreover, that the amounts of protein and RNA which are formed and incorporated into the chromosomes of plants may be varied by the addition of extraneous solutes, such as phosphates.

## VIII. Chromosome Puffs, Lampbrush Chromosomes, and the Y Chromosome in *Drosophila*

The production of puffs on the giant chromosomes of larvae in various species of *Drosophila* and in the midge *Chironomus* has been reviewed by Beermann (1967; see also Beermann and Clever, 1964).

The giant chromosomes occur in a variety of metabolically active cells, such as those in the salivary glands, intestines, and Malpighian tubules. They are unusual in that the cells no longer divide, but they grow in size and the chromosomes undergo repeated DNA replications. As a result, they grow in thickness as well as in length and show a characteristic and specific pattern of darkly staining bands, which vary in thickness, separated by lightly staining interbands. The bands are rich in DNA and histones, and recent evidence suggests that these substances are also present in the interbands, though in much smaller amounts than are found in the bands. The bands are also called "chromomeres." Although, in the past, there was a tendency to regard the bands as visible manifestations of genes, Crick (1971) has recently suggested that their function may be regulatory, while the coding genes may be located in the interbands.

The pattern of bands and interbands has been found to be constant in different tissues but this constancy does not apply to the production of puffs. These are localized protuberances on the giant chromosome, which are thought to arise by an unfolding of the DNA in a band. The puffs are active producers of RNA, whose base composition suggests that it is the product of only one DNA strand. Large puffs are known as Balbiani rings.

Beermann and his collaborators have shown that the formation of puffs on particular bands is specific for certain tissues and varies with different stages of development. Changes in puff formation in the salivary glands during metamorphosis of *Chironomus* larvae were studied in detail. Although some puffs did not show any relation to metamorphosis, others appeared at definite times after metamorphosis had begun. Other puffs again, although present in larvae of all ages, were particularly large during metamorphosis. The hormone ecdysone was found to control not only metamorphosis but also the formation of the correct puffing pattern during metamorphosis. Puffing patterns may also be induced by temperature shocks (Ashburner, 1970).

Although it was originally assumed that the puffs might be the result of differential gene activity during development, it now seems more likely that they are due to the activity of much larger chromosomal regions, which are especially active in RNA synthesis during certain stages of development. These puffs, therefore, demonstrate that specific chromosomal regions may regulate the amounts of synthetic activity in particular stages of development.

Lampbrush chromosomes are another example of chromosomal DNA being present in a greatly extended state which is highly active in RNA synthesis during a particular stage of development. They have been defined as germline chromosomes with laterally projecting loops (Callan, 1963). Typical lampbrush chromosomes are found in the oocytes of animals with yolky eggs and have been studied in most detail in urodele amphibia such as newts (*Triturus*), which have very large chromosomes; lampbrush chromosomes have also been described in other vertebrates and in some invertebrates (Hess, 1970). Lampbrush chromosomes are in the diplotene stage of the first prophase of meiosis. Each chromosome, therefore, represents a bivalent with chiasmata. In the twelve bivalents of *Triturus cristatus*, about 5000 pairs of individual loops have been identified; each pair was found to be associated with a particular chromomere.

It appears that only a small proportion of the dry mass of lampbrush chromosomes is due to DNA, the rest being accounted for by RNA and protein (Izawa *et al.*, 1963). In *Xenopus laevis* it has been estimated that over 90% of the RNA is ribosomal. In order to produce these great amounts, the nucleolar DNA is increased, or amplified, in oocytes (Brown and Dawid, 1968). This DNA appears as the satellite fraction in calcium chloride gradients (Chapter 4, Section VI). Although the proportion of messenger RNA capable of protein synthesis is only 2% in these nuclei, this still amounts to an increase of several thousand times compared with the RNA contents of somatic chromosomes (Davidson *et al.*, 1966).

Although less spectacular than in amphibian oocytes, lampbrushlike chromosomes have also been demonstrated in the spermatocytes of several animal species. Recent findings on the Y chromosomes in various species of *Drosophila* (Hess, 1970) are of special interest in the context of the activity of sex chromosomes.

As mentioned before (Chapter 1, Section VII), the Y chromosome in *Drosophila melanogaster* is not required for male development, but its presence is necessary to ensure male fertility. In the absence of a Y chromosome, sperm are formed which are immotile, even if they carry an X chromosome. It is clear, therefore, that the Y chromosome must function at a stage before it has disjoined from the X chromosome. It has been shown (Hess and Meyer, 1968; Hess, 1970) that during the first prophase of meiosis in the spermatocytes of many species of *Drosophila,* the Y chromosome forms lampbrushlike structures. These are particularly large in *D. hydei* and appear as a pair of compact threads arising in the vicinity of the nucleolus. The threads as well as certain other structures– collectively known as "loops"—are absent in XO males and are present in a double dose in XYY males (Fig. 7.5). The authors assume that these structures develop by an unfolding of the DNA from certain regions of the Y chromosome during the growth period of the spermatocyte. The structures are highly active in RNA synthesis, which is temporarily blocked both by actinomycin and by X rays.

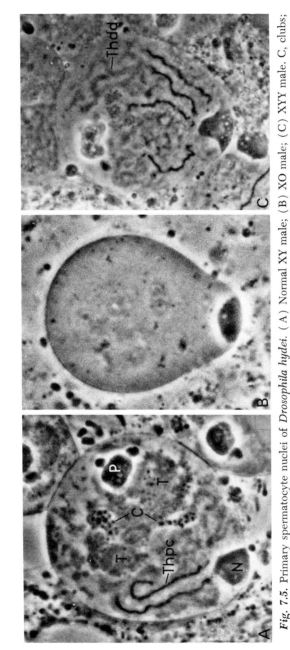

*Fig. 7.5.* Primary spermatocyte nuclei of *Drosophila hydei*. (A) Normal XY male; (B) XO male; (C) XYY male. C, clubs; N, nucleolus; P, pseudonucleolus. T, tubular ribbons; Thdd, Thpc, distal diffuse and proximal compact sections of threads, respectively. Phase contrast photographs of living nuclei. (From Hess and Meyer, 1968; photographs contributed by O. Hess.)

Results with structurally aberrant Y chromosomes indicated that the threads and other special structures visible in the spermatocytes originate in a few regions situated on the long as well as on the short arm of the Y chromosome. If a Y chromosome lacks any of these regions, spermiogenesis stops at an intermediate stage and the sperm remains immotile. It is of interest to note that the DNA in these regions does not seem to code for any protein necessary for an organelle in the sperm. However, the various loop-forming regions on the Y chromosome control the growth of the sperm; males with two Y chromosomes form sperm of approximately twice the normal length while various duplications cause intermediate elongations of the sperm.

## IX. Changes in Karyotypes and Mitotic Rates

When two organisms differ in their chromosome constitution, a difference in growth rate is frequently apparent.

Polyploidy presents a special problem because the duplication of chromosome number results in an increase in nuclear size. In certain plants and invertebrates, polyploidy is, indeed, associated with increased size (Beatty, 1957). In amphibia, however, polyploid animals are of about normal size, in spite of having larger cells (Fankhauser, 1945); this suggests a reduced rate of growth.

In mammals, polyploidy has a lethal effect. Triploid fetuses are known in mice (Fischberg and Beatty, 1951), rats (Piko and Bomsel-Helmreich, 1960), rabbits (Bomsel-Helmreich, 1970), and in man (Carr, 1971a), in whom exceptionally triploid fetuses have been known to survive birth. In spite of an increase in cell sizes, the triploid fetuses were below normal in size in all four species, thus suggesting a considerable decrease in the rate of cell proliferation.

Van't Hof (1965) has presented evidence that in various species of plants those with a higher DNA content have a longer mitotic cycle time. This relationship, however, does not seem to hold for autotetraploidy, in which mitotic cycle times are similar to the diploid ones (Troy and Wimber, 1968; Yang and Dodson, 1970). Goldfeder (1965) showed that spindle tumors, which contain a

large proportion of polyploid cells, require a longer time to synthesize DNA than epithelial tumors composed predominantly of
diploid cells. Similarly, ascites tumor cells with tetraploid karyotypes have been shown to require about twice as long to complete
DNA synthesis as diploid ascites cells (Lennartz *et al.*, 1966). In
a fibroblast culture originating from a patient who was a chimaera
of diploid and triploid cells, Mittwoch and Delhanty (1972) presented evidence suggesting an inhibition of DNA synthesis and
mitosis in the triploid cells when compared with the diploid ones.

From polyploidy we may turn to trisomic conditions. The most
common autosomal trisomy in man, mongolism, is associated with
a reduced birth weight and delayed skeletal development. Hall
(1965, 1966) suggested that the clinical abnormalities seen in mongolism—as well as those in the rarer autosomal trisomic conditions,
trisomies 13 and 18—may be due to a slowing down of cell multiplication as a result of the presence of the additional chromosomes.
Hall also pointed out that the pathological conditions due to the
presence of an extra chromosome may also be produced by environmental agents such as drug therapy. Trisomies 13 and 18, like
mongolism, are also associated with reduced birth weights, and
Naye (1967) has shown that all three contain a subnormal number
of cells in a variety of organs at birth. Preliminary evidence on
cultured fibroblasts suggested that those with a mongol karyotype
may be slower to synthesize DNA than normal fibroblasts (Mittwoch 1967c). Support for this suggestion has come from the finding
by Kaback and Bernstein (1970) that the rate of DNA synthesis
in fibroblasts from patients with mongolism is slowed down; while
Mellman *et al.* (1970) reported that the maximum rate of DNA
synthesis in lymphocyte cultures from mongols was less than in
control cultures.

The most exhaustive study of trisomy was carried out by Blakeslee on a flowering plant, *Datura stramonium*, the jimsonweed (see
Avery *et al.*, 1959). This species has twelve pairs of chromosomes
and all twelve trisomic types have been produced and analyzed.
Each one was found to be morphologically distinct and differing
from each other and from the normal type by such features as
the habit of the plants, the shapes and sizes of the capsules, and
the morphology of the spines. Each trisomic type also showed

slower growth than the plants with normal karyotypes. The last fact is particularly difficult to reconcile on any theory postulating that growth is controlled by specific genes, since it would require the assumption that each chromosome in *Datura* carries an excess of genes for slow growth. The postulate, that the addition of any chromosome to the normal karyotype slows down the normal rate of cell proliferation would seem to be more realistic.

The effect of supernumerary B chromosomes in increasing the time required to complete the mitotic cycle in rye plants (Ayonoadu and Rees, 1968) has already been mentioned (Chapter 4, Section VII). B chromosomes differ from trisomic ones in not being part of the normal karyotype and in showing heterochromatic properties. It appears that any type of chromosome which is present in addition to the normal complement may have the effect of prolonging the mitotic cycle time.

In apparent contrast to these findings is an earlier report by Darlington and Thomas (1941) that B chromosomes in *Sorghum* plants increased the number of mitoses in pollen mother cells, even though these chromosomes were lost in other cells. This finding suggests the possibilities that in certain circumstances the presence of a given chromosome or chromosomal region may increase mitotic rates in the cells of a particular organ.

In the context of chromosome number and its relation to growth, mention should be made of the relationship of different sex chromosome constitutions and the number of dermal ridges in man. Dermal ridges, which are present, for instance, on the tips of the fingers, are a countable characteristic, which, unlike most, is not affected by environmental influences throughout the life of an individual. They are laid down in the fetus during the third and fourth month of pregnancy and thereafter do not change (Holt, 1968). It is likely that the ridge count is related to the growth attained by the fetus during the fourth month of pregnancy. The total ridge count, which is obtained by adding the counts on the ten fingers together, shows a high correlation between parents and children. The mean total ridge count, however, is somewhat lower in females than in males and different values are also found in patients with abnormal numbers of sex chromosomes (Penrose, 1967, 1969; Polani, 1969). It is evident from their data that the mean total ridge count decreases steadily with the addition of more sex chromosomes to the karyotype, but that the X chromosome has a more marked effect in this respect than the Y chromosome. It is of interest that the count for patients with Turner's syndrome (45, X) is even higher than those for normal males. This might be connected with the finding, admittedly preliminary at present, that the mitotic cycle time of cultured cells with 45, X karyotypes was shorter than of cells with 46, XX or 46, XY karyotypes (Barlow, 1972). If this finding were confirmed, it would imply that the low birth weight and small stature of patients with Turner's syndrome would be indirect effects of the abnormal karyotype, resulting from faulty embryogenesis.

The ridge count data suggest that in many types of human cells, the addition of the Y chromosome to the karyotype may result in a slight decrease in the rate of growth, while the addition of an X chromosome would result in a more marked slowing down of growth. This general scheme is likely to be modified in cells of the gonadal rudiments; their growth will be further discussed in Section XII (this chapter).

## X. The Hierarchy of Genetic Elements

The many chromosome aberrations which have come to light in the human species have led to a reappraisal of the question of the relationship between abnormal chromosome constitutions and the resulting maldevelopment of the phenotype. These results seem to be leading to the conclusion that many such abnormalities cannot be assigned to the action of specific gene loci.

There has, of course, been no lack of attempts to do so. For instance, when it was found that patients with mongolism have an increased level of leukocyte alkaline phosphatase in their neutrophil polymorphonuclear leukocytes (King et al., 1962), a number of investigators have postulated the existence of a gene for leukocyte alkaline phosphatase production on chromosome 21, which, when present in triplicate, was assumed to produce increased amounts of the enzyme (see Hamerton, 1971b). However, Hsia et al. (1964) found increased levels of glucose-6-phosphate dehydrogenase, which is known to be due to a gene located on the X chromosome, in patients with mongolism. Weber *et al.* (1965) reported that the leukocyte alkaline phosphatase level is raised in patients with Klinefelter's syndrome. Apparently, abnormal enzyme levels may be produced by a variety of chromosomal abnormalities, which affect the growth and maturation of cells. In trisomies 13 and 18 unsuccessful attempts have also been made to correlate phenotypical abnormalities with genes borne on the chromosomes in question. It is, indeed, remarkable how much overlap is seen in the phenotypes produced by different chromosomal abnormalities (e.g., Taylor, 1968).

Edwards (1970) has recently drawn attention to the hierarchical organization of the karyotype, originally developed by Goldschmidt (1955), from the diploid chromosome set down to the nucleotide (Fig. 7.6). In view of the frequently made assertion of excellent agreement between cytological and genetic data, it is perhaps ironical that only changes in the upper part of the diagram are visible under the microscope, while the only type of change whose pheno-

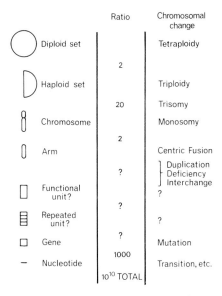

*Fig. 7.6.* The hierarchical organization of genetic elements. (From Edwards, 1970.)

typic effect is understood is that of the penultimate unit, the gene. Regarding changes in chromosomal units other than genes, not only do we lack any generally accepted theories of their effects, but since only the grossest are visible, we do not even know of the existence of a large proportion of them. Yet there is little doubt that chromosomal changes which are intermediate in size between those which are easily visible, on the one hand, and genuine gene mutations, on the other, do in fact occur. As pointed out by Edwards (1970), human chromosomes are very variable and, if parents are examined alongside their children, chromosome arms of unusual length, or with peculiar satellites, can almost always be seen to have been transmitted unchanged. It now seems certain that by means of recently discovered techniques of selective staining, small chromosomal changes, which could not be distinguished with certainty in the past, will become visible under the microscope. Thus, Evans *et al.* (1971) described seven polymorphic regions on human chromosomes, which could be distinguished by means of quinacrine fluorescence and acetic–saline–Giemsa techniques. Once such chromosomal variants are established, an attempt at evaluating their effect on the phenotype can be made. Barr and Ellinson (1971) have recently reported extensive differences in fluorescent staining patterns between *Drosophila melanogaster* and *D. simulans,* whose chromosomes appear very similar by conventional staining techniques.

Enlarged satellites were found by Goodman *et al.* (1968) in the chromosomes of 20 out of 36 baseball players, whose height varied between 180 and 208 cm. The large satellites were seen mainly in the D group chromosomes. It is conceivable that this finding may correlate with an earlier report by Tjio *et al.* (1960) of describing the presence of enlarged satellites in patients with Marfan's syndrome. This is a heritable disease of connective tissue, with skeletal and other abnormalities (McKusick, 1960); the patients tend to be tall. Although the association between enlarged satellites and Marfan's syndrome is still in doubt (Handmaker, 1963), it would seem that Handmaker's own findings do not disprove this possibility.

Satellites have the appearance of heterochromatin (Chapter 4, Section V). The effects on development by changes in the heterochromatic regions of chromosomes are likely to be of a quantitative nature, affecting growth rates, and these can be modified by environmental agents. It is likely that many phenotypic differences, which are partly determined by heredity, and for which the existence of many genes has been postulated, are in fact caused by small changes in the pattern of the chromosomes, which were not detectable by the cytological techniques available in the past. Those phenotypic effects, which show different incidences in males and females, are of special interest in the present context, since they may be assumed to result from the interaction between a small chromosomal change and the established major dimorphism of the sex chromosomes. The more viable changes of this type are likely to be due to repatterning in heterochromatic regions, since similar changes in euchromatin would be expected to have more marked and, therefore, deleterious effects.

As a general rule, one might expect chromosomal changes to have an effect on growth and, unless this is very marked, to be modifiable by environmental agents. Conversely, those genetic effects, which are found to be variable in nonspecifically different environments, are the most likely to be caused by chromosomal changes rather than by gene mutations.

Although seemingly obvious, the hierarchical organization of the chromosomal material has so far been largely neglected. Admittedly, the fact that changes in the chromosomes have played a large part in the evolution of animals (White, 1954) and of plants (Darlington, 1958) has been recognized for many years, but this knowledge has had virtually no effect on studies of population genetics and evolution theory. There has been a tendency among geneticists, and even among cytogeneticists, to regard all hereditary characters as due to genes and the chromosomes as mere vessels which carry the genes. Consequently, if any such characters do not conform to Mendelian ratios, additional concepts of penetrance, expressivity, or multifactorial inheritance have been postulated to enable the facts of heredity and development to fit the gene concept.

Goldschmidt (1954, 1955) was an outstanding opponent of the classical concept of the gene. He conceived the chromosome as a hierarchical structure without "corpuscular" genes, in which the smallest units, the loci, may interact in different combinations to make up fields of higher order, as shown in the following scheme:

Loci                    1  2  3  4  5  6  7  8  9  10  11  12 → $n$

Segments

Fields 2nd order

Fields 3rd order

Chromosomes

This hierarchical organization was thought to have increased in complexity during the course of evolution as a result of "repatterning" of the chromosomes.

This scheme of chromosomal organization was conceived before the advent of molecular genetics, when the physical nature of the gene was entirely unknown; its virtually complete neglect in subsequent years, when the genetic code was discovered, is hardly surprising. By now, however, it has become clear that the genes are not the only components of the chromosomes of higher organisms, since they contain far more DNA than can be accounted for by the sum total of genes coding for proteins and they also contain varying amounts of non-DNA components, mainly proteins. Now that the nature of the gene is understood, the search to understand the structure and function of the nongenic parts of chromosomes has become a major task in biology.

## XI. Possible Functions of Nongenic DNA

The evolution of higher organisms has been accompanied by a massive increase in the DNA contents of their cell nuclei. As illustrated in Fig. 7.7 mammals have almost a thousand times more DNA per cell than do bacteria.

The increase in DNA which accompanied the evolution of vertebrates has been documented by Ohno and collaborators (Atkin et al., 1965, Ohno and Atkin, 1966; Atkin and Ohno, 1967; Ohno 1970). Among primitive chordates, the sea squirt, Ciona intestinalis, was found to have a DNA value of about 6% of that of

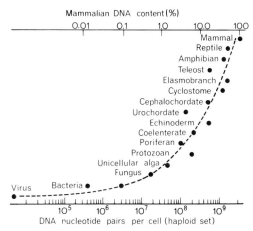

**Fig. 7.7.** Diagram illustrating the increase in DNA contents per haploid cell during the course of evolution. (From Britten and Davidson, 1969.)

the human species; the corresponding figures for the lancelet, *Amphioxus lanceolatus*, were 17%, for the brook lamprey, *Lampreta planeri*, 38%, and for the hagfish, *Eptatretus stoutii*, 78%.

Among teleost fish, the swordtail (*Xiphophorus helleri*) the hornyhead turbot (*Pleuronichthys verticalis*) and the fantail sole (*Xystreurys liolepis*) had values of 20% of that found in man; corresponding values for the green sunfish, *Lepomis cyanellus*, and the discus fish, *Symphysodon aequifasciata*, were 30–35%; for the goldfish, *Carassius auratus*, about 50%; the rainbow trout, *Salmo iridens*, 80%; while a specimen of South American lungfish, *Lepidosiren paradoxa*, had a DNA value which was thirty-five times the human one!

Among amphibians, urodeles, including newts and salamanders, had high DNA values which varied between seven and twenty-eight times that of man, while the DNA values of frogs and toads (Anura) varied between 40 and 250% of the human value.

Placental mammals were regarded as a uniform group, even though some variation between species undoubtedly occurs. Birds were found to have roughly 50% the mammalian DNA contents, while reptiles fell into two classes: six representatives of the order Squamata, including snakes and lizards, had DNA values from

60 to 67% of mammalian ones; while three species belonging to
the order, Crocodilia and Chelonia (turtles) had DNA values between 80 and 89% of those of mammals.

According to Rasch *et al.* (1971), the DNA contents of diploid
nuclei of *Drosophila melanogaster* is about one-seventh of that
of the chicken, which would mean that the DNA contents in *Drosophila* is roughly 7% of that of mammals.

These findings show two trends. The first is that evolution has
been accompanied by a steady increase in DNA, and, second, that
this process seems to have continued to apparently excessive degrees in certain groups, such as the urodele amphibia and the
South American lungfish, *Lepidosiren paradoxa.* This species has
38 metacentric (and submetacentric) chromosomes, all of which
are very large. By contrast, the swordtail, *Xiphophorus helleri,* has
48 small acrocentric chromosomes, while the primitive chordate,
*Ciona intestinalis* has 28 small acrocentric chromosomes. It is clear
that evolution must have come about both by an increase in size
of chromosome arms as well as by an increase in chromosome
numbers. During this process, genes which were originally present
in single dose came to be present in multiple doses.

Vogel (1964) suggested that the major part of the DNA of higher
organisms had a regulatory function and is connected with structural genes in operons. Structural genes, which alone code for poly-
peptide chains to synthesize enzymes and other proteins might
make up only about 1% of the total DNA. Models of gene regulation
have been constructed by Britten and Davidson (1969) and by
Tomkins *et al.* (1969). In the model by Crick (1971), the chromosomes are divided into bands and interbands, as seen in the giant
chromosomes of *Drosophila.* The bands, which contain the major
part of the DNA, are thought to consist of single-stranded DNA,
which controls the minor fraction of double-stranded DNA situated
in the interbands, which alone codes for proteins.

Ohno (1970) has put forward the view that the duplication of
genes has resulted in easing selection pressure by allowing mutations to occur at existing loci, while the duplicate genes could
carry on their original function.

It seems unlikely that any of the theories proposed so far provide
the entire explanation of the effect of the massive increase of DNA

on the development of organisms. Regarding the amounts of enzyme produced by a given gene, the evidence from human genetics
seems to suggest that this can vary to a considerable extent without
untoward effect. For instance, patients suffering from phenylketonuria or other inborn errors of metabolism are homozygous
for a gene which produces an ineffective enzyme. In heterozygous
carriers of this condition, the effective enzyme is usually produced
in half the normal amount (Harris, 1970). However, it requires
a considerable amount of experimentation to detect such a decrease
in the amount of enzyme and, even then, the enzyme levels found
in heterozygous carriers tend to overlap with those of homozygous
normal persons. The heterozygous carriers do not seem to be clini-
cally or otherwise affected, so that for practical purposes the inborn
errors of metabolism rank as "recessive." It would seem unlikely
that perhaps 99% of the DNA should be involved in this type of
apparently not very critical control.

A glance at Fig. 7.7 suggests that the increase of DNA from
virus to mammal is also associated with a general increase in the
size of the organism, the causes of which must be numerous. First,
the increased amount of DNA itself will bring about an increase
in the size of the nucleus and, consequently, the cell, and this
increased cell size will be further enhanced by increasing amounts
of protein and RNA which become associated with the larger chromosomes. As the chromosomal material becomes enlarged, one may
expect mitotic cycle times, in general, to become prolonged, thus
prolonging the period of development. Nevertheless, these chromosomes may still be capable of more rapid cell division in certain
stages of development, and this allows for the possibility of differential growth, which is such a characteristic feature of complex
organogenesis.

Although the subject of chromosome size in relation to the time
taken for overall division is as yet little explored, we may assume
that as originally unique gene loci became multiplied, many of
them ceasing to function in their original capacity and becoming
part of the heterochromatin. In this condition, the DNA sequences
may be assumed to have changed rapidly by mutation, while larger
stretches of the chromosomes have become rearranged in various
ways, e.g., by duplication, deletion, and translocation thus giving

rise to the repatterning, which Goldschmidt described, and which is such a striking feature when the chromosomes of different species are compared (White, 1954).

The sex chromosomes present a special problem of maintaining a chromosome dimorphism within the same species. Highly evolved sex chromosomes must have originated from a pair of chromosomes which were originally equal but progressively became different from each other (Darlington, 1958). During this process there has been a tendency for the Y chromosome, which is present only in one sex, to become largely, if not entirely, heterochromatic, and this process may be followed by the loss of part of the chromosome. This is likely to have happened in mammals, where the Y chromosome has retained a region which is active in testicular differentiation. The Y chromosome in certain species of *Drosophila* provides an example of a heterochromatic chromosome which is active in RNA synthesis during meiosis (Hess, 1970). That the mammalian Y chromosome may be active in RNA synthesis during certain mitotic stages of development would thus seem to be a possibility.

The X chromosome has retained its euchromatic components, but one of its members becomes facultatively heterochromatic in female mammals. Facultative heterochromatinization of the X chromosome does not seem to occur in birds, where both X chromosomes of the male are euchromatic (Schmid, 1962); the same applies to snakes (Ray-Chaudhuri *et al.*, 1970).

## XII. Sex Differentiation and Growth

In spite of the presence of special sex chromosome mechanisms, it has generally been assumed that maleness and femaleness is due to allelic differences at particular gene loci. The classical theory of sex determination, which was based on the fruit fly, *Drosophila melanogaster,* postulated large numbers of male- and female-determining genes, the female-determining ones being present on the X chromosome and the male-determining genes situated on the autosomes (Bridges, 1939; see also Chapter 1, Sections VII and IX). According to this theory, the importance of different numbers of X chromosomes in *Drosophila* lies in the fact that they carry large numbers of female-determining genes, which will either outnumber the male-determining ones on the autosomes, or be out-numbered by them.

The large number of sex-determining genes were required be-cause, first, the Y chromosome in *Drosophila* has no male-determin-ing function, and, second, the female-determining part of the X chromosome is not concentrated on any particular locus. In mam-mals, the formal scheme of sex determination appears to be simpler. The Y chromosome causes the development of testes which induces the male phenotype, while absence of a Y chromosome results in the absence of testes and the development of the female pheno-type. Recently several authors have proposed theories of sex deter-mination which are not based on a preponderance of male- and female-determining genes (Hamerton, 1968, 1971b; Boczkowski, 1971; Ohno, 1971; Ohno *et al.*, 1971). In Hamerton's scheme, the mammalian X chromosome is thought to carry a male-determining gene, while the Y chromosome carries a controlling center which activates this gene into producing a male inducer substance, which, in turn, stimulates the medullary component of the gonad to de-velop into a testis. Originally another female-determining gene causing ovarian development was also postulated to be on the X chromosome, but this was subsequently abandoned as unneces-sary (Hamerton, 1971b, addendum). It would seem, however, that the male-determining gene is equally unnecessary, nor if there were one, would there be any evidence for its location. All we know is that the Y chromosome is the primary sex determiner.

The theory of Boczkowski contains three postulates. The first is that the mammalian gonad will differentiate into a testis, unless it is inhibited from doing so; second, that the specific inhibiting factor is produced by a structural gene probably located on the X chromosome; third, that this gene is, in turn, rendered inactive by a repressor substance produced by a structural gene located on the Y chromosome. Once again, the only part of the theory which is supported by factual data is the activity of the Y chromo-some. The author bases his first postulate on the findings of au-tonomous differentiation of transplanted testicular rudiments; this, however, could equally be due to the activity of the Y chromosome in these rudiments.

Ohno's theory is concerned only with the differentiation of the secondary sexual characters. It proposes that maleness and female-ness are the induced and noninduced states, respectively, of a single regulatory system, in which the inducer is testosterone. When this is present, male development is thought to be induced by the action of a single regulator gene, which is situated on the X chromosome. This step in the argument is based on findings in the syndrome of testicular feminization (Chapter 6, Section VI). The condition, in which female external genitalia are present in spite of XY sex chromosomes and testes, occurs in man as well as the mouse. In the latter species, there is good evidence that the condition is due to a gene borne on the X chromosome and this makes it likely that the same mode of transmission operates in man. In this syn-drome, the testes produce testosterone, but the target tissues do not react to it; Ohno and collaborators, using proximal kidney tub-ule cells, which are developmentally related to Wolffian duct cells, found that these cells in normal males and females responded to testosterone by producing kidney alcohol dehydrogenase and $\beta$-glu-curonidase, whereas no enzyme production took place in the kidney cells of mice with testicular feminization. On the basis of these findings it was proposed that the normal allele of the testicular feminization gene produces a protein in the target cells of testoster-one, which, in the absence of the testosterone acts as a repressor of the translation of certain enzymes, but when bound with testos-terone activates RNA polymerase to produce certain sets of enzymes.

While the claim that "a single gene product mediates the entire testosterone-induced response of target cells" (Ohno, 1971) may be valid, the statement that "Whether a mammalian embryo is to manifest the male phenotype or the female phenotype is solely dependent upon the presence or absence of testosterone" (Ohno et al., 1971) is clearly a gross oversimplification. Not only is the primary sex difference—the nature of the gonads—unaffected by testosterone, but this hormone when administered exogenously also seems to be incapable of causing the Mullerian ducts to regress. Therefore, in order to account for the absent uterus in animals or patients with testicular feminization, it is necessary to postulate the existence of "factor X," acting alike in normal males as in testicular feminization. In addition, however, patients with this syndrome show a large number of other differences compared with normal females, including one of stature, the distribution of body hair, and an increased tendency of the gonads to become malignant (Hauser, 1963b, Polani, 1970). Therefore, in spite of outward appearances, the gene mutation in testicular feminization does not render a chromosomal male into a phenotypic female, but merely into a male pseudohermaphrodite.

A somewhat analogous condition has long been known in *Drosophila melanogaster*. Sturtevant (1945) described a gene located on the third chromosome, which transformed XX flies into males, complete with sex combs, male-colored abdomens, and male external and internal genitalia. The testes, however, are small and the flies are sterile. The transformed males resemble females, in size, in rate of development, and the expression of certain sex-limited characters.

Examples such as these illustrate that in spite of the existence of apparently simple switch mechanisms, the problem of sex differentiation in higher organisms cannot be solved by recourse to models borrowed from bacteria. Indeed both the evolutionary evidence, and the chromosomal evidence argue against a simple gene difference.

When we compare the sex-determining mechanisms in different classes of vertebrates, we are struck by two findings. One is that as we descend in the evolutionary scale, gonochorism, or the separation of the sexes in different individuals, tends to be less strictly observed than it is in mammals and birds. Even though habitually functional hermaphroditism, as in the fish, *Rivulus marmoratus* (Harrington, 1968), is the exception, we have seen (Chapter 5, Section IV) that developmental hermaphroditism is not uncommon in fish. Moreover, different species may show either protogynous hermaphroditism (female phase develops first) or protandrous hermaphroditism (male phase develops first). This is evidence not only that the development of one or other sex is related to the stage of development of the individual, but also that, in different species, the more juvenile phase can be either female or male.

In relatively early stages of evolution it may still be possible to reverse the predetermined course of sex differentiation by means of rather nonspecific changes in the environment, as in the case of chromosomal male Mexican *Xiphophorus maculatus* developing as female as a result of X irradiation (Anders *et al.*, 1969a) or the normally hermaphrodite *Rivulus marmoratus* developing as males in cold temperatures (Harrington, 1968). Examples such as these render the view held at the turn of the century, that sex is determined by the combined effect of external conditions (Wilson, 1896) more intelligible. Although it is now clear that data purporting to show such an effect in, for instance, mammals were based on inadequate statistics, the evidence suggests that subsequent attempts made to explain sex in terms of Mendelian genes were equally misplaced. A more likely assumption is that sex differentiation originated in differential growth mediated by environmental agents and has subsequently come under the control of special chromosomal regions by a process of genetic assimilation.

The fact that both protogynous and protandrous hermaphroditism are to be found may possibly be connected with the other striking feature, the apparently haphazard distribution of male and female heterogamety in different classes of vertebrates. As we saw in Chapter 5, both conditions occur in fish and can even be found within the same species, for example, the platyfish, *Xiphophorus maculatus;* both conditions also occur side by side in amphibia. In mammals and in birds, the systems have become stabilized, one in each class. Evidence has been presented that in mammalian embryos, the testicular gonadal rudiment grows faster than the ovarian rudiment, while the opposite situation seems to prevail in birds. Furthermore, the embryonic testis in mammals soon becomes a secretor of androgens, which then masculinize the reproductive tract, whereas the ovary plays no major part in the sexual development of the embryo. By contrast, in birds, the embryonic ovary is the dominant gonad, which secretes estrogens and these, in turn, feminize the reproductive tract. This contrasting state of affairs between birds and mammals would seem to be further evidence against Ohno's suggestion that, in mammals, maleness and femaleness are the induced and noninduced state of the same regulatory system. It seems more likely that, beginning at somewhat different starting points in the bipotential gonad, an accelerated growth rate leads to early cortical development in birds and early medullary development in mammals.

It has already been suggested that the early testicular differentiation of mammalian embryos may be necessary to allow masculine
differentiation to proceed within the female hormonal environment
of the uterus. Indeed, the strong masculinizing power of the mammalian Y chromosome may be regarded as having arisen in response
to this difficult situation, in which the male mammalian embryo
finds itself. By contrast, the male avian embryo does not face this
problem and so the female embryo is free from an early age to
prepare for its future role as egg layer.

The highly evolved process of sex differentiation in birds and
mammals has led to the unusual situation of an intraspecific chromosomal dimorphism of an order of magnitude which is easily
detectable under the light microscope. Allelic differences between
Mendelian genes are inferred from their mode of segregation
in different generations but cannot be seen under the microscope.
As a corollary, we may assume that where obvious differences
in the chromosomes are visible, the underlying effect on development cannot be accounted for in terms of genes. On the other
hand, evidence is slowly accumulating that different chromosome
constitutions may affect the rates at which cells divide, including
the different stages of mitosis and meiosis.

There are several well-known correlations between sex and
growth. In the human species, the growth rate and average size
differ in males and females; males are already larger than females
at birth (Tanner, 1962). It is, of course, not known so far whether
this difference is a direct effect of the difference in chromosome
constitution, or whether it is a secondary effect, for instance, as
a result of a hormonal difference.

The gametes themselves exhibit quantitative differences in a particularly striking way. Eggs and sperm are homologous cells, which
originate from the same type of primordial germ cells; both undergo
a series of mitoses, followed by two meiotic divisions. Yet in spite
of the similarity of their backgrounds, the volume of mammalian
egg cells is many thousand times greater than that of the sperm.
The difference in numbers is even greater. Mammalian ova are
used sparingly, whereas sperm are produced in abundance. Thus,
a woman produces less than five hundred ova throughout her life
time, while a human male uses up about three hundred million

spermatozoa in a single ejaculation. Last, there is a great difference
in the time taken over various stages of meiosis in male and female
germ cells. Female germ cells spend a very long time in the prophase stage of the first meiotic division. This may vary according
to species from a few months to many years; in the human female,
the time taken over the prophase of the first meiotic division is
between 12 and 50 years. The duration of the first meiotic prophase
in the human male has been estimated as 23 days and that of the
entire process of spermatogenesis as 74 days (see Courot *et al.*,
1970).

In birds, the differences between male and female germ cells
are basically similar to those in mammals, except that the difference
in size between egg and sperm is, of course, vastly greater.

In addition to gametes, the gonads also produce sex hormones
and these, too, show some interesting quantitative differences between males and females. All naturally occurring sex hormones
are steroid, which are customarily divided into androgens, or masculinizing hormones, estrogens, or feminizing hormones, and gestagens, which are required to maintain pregnancies in mammals
(Corner, 1947; Villee, 1961; Zander and Henning, 1963; Bloch,
1967). However, in spite of this classification, a particular group
of sex hormones is not confined to one sex. The main hormones
produced by the human ovary are estrogens and gestagens but
a small amount of androstenedione, an androgen, has also been
isolated. Androgens are also produced by the adrenal cortex in
both sexes. Testosterone is the main hormone secreted by the human testis but, in addition to androgens, the testes also produce
a small amount of estrogen. Indeed, shortly after the discovery
of estrogens in the 1920's, it was found that large quantities of
these hormones could be obtained from the urine of stallions
(Corner, 1947). It is clear that both testes and ovaries are secre-
tors of androgens, estrogens, and gestagens, but that the quantitative relationships of these hormones vary according to sex, species,
age, and season.

There is no sharp division between estrogens, androgens, and
gestagens. Some give both androgenic and estrogenic effects and
may even affect the uterine lining, although only in large doses.
It has often been reported that androgens, given in excessive doses,

promote the development of the female genital tract, and it is possible that estrogens may sometimes induce the growth of male structures. This so-called "paradoxical effect" is observed only with abnormally large doses of sex hormones (Corner, 1947).

The exact way in which steroid hormones produce their effects is not yet known, but it is generally assumed that an essential part of the mechanism is the binding of the steroid to a protein in the target organ (Villee, 1961). The difference in different target organs, e.g., those that respond to androgens and those that respond to estrogens, is assumed to be due to a difference in the type of proteins present in these organs. The evidence suggests, however, that the type of steroid hormone produced is not itself dependent on a specific protein and that both the production and the effects of steroid hormones are regulated by quantitative variables.

Hormones are known to affect the duration of the mitotic cycle (Epifanova, 1971). Thus, estrone administered to ovariectomized mice has been shown to shorten the mitotic cycle time of cells in the uterine epithelium and may also induce cells to enter the cycle. The possibility exists, therefore, that, following gonadal differentiation under the direct influence of the sex chromosomes, the steroid hormones may affect the differentiation of secondary sexual characteristics by controlling the growth rate of target organs.

A general neglect by scientists applies not only to differences in chromosome constitution but also to differences in growth. Many years ago, Haldane (1927) wrote as follows: "The most obvious difference between different animals are differences of size, but for some reason the zoologists have paid singularly little attention to them. In a large textbook of zoology before me I find no indication that the eagle is larger than the sparrow, or the hippopotamus bigger than the hare, through some grudging admissions are made in the case of the mouse and the whale. But yet it is easy to show that a hare could not be as large as a hippopotamus, or a whale as small as a herring. For every type of animal there is a most convenient size, and a large change in size inevitably carries with it a change of form."

The evolution of sex differentiation shares two fundamental features with the evolution of species. In both cases the chromosomes become different and the organisms come to differ in size and

shape. However, the third feature is antithetic, for whereas specia-
tion implies genetic isolation, sex differentiation leads to obligatory
interbreeding. Accordingly, only one pair of chromosomes has been
modified to provide two chromosome constitutions which bring
about differences in the rate of development. The rest of the
chromosomes remain the same in both sexes, thus ensuring perfect
coordination in mitosis and in meiosis.

# Bibliography

Aida, T. (1921). On the inheritance of color in a fresh-water fish, *Aplocheilus latipes* Temmick and Schlegel, with special reference to sex-linked inheritance. *Genetics* **6**, 554–573.

Aida, T. (1936). Sex reversal in *Aplocheilus latipes* and a new explanation of sex differentiation. *Genetics* **21**, 136–153.

Alexander, G., and Williams, D. (1964). Ovine freemartins. *Nature* (*London*) **201**, 1296–1298.

Alfert, M., Bern, H. A., and Kahn, R. H. (1955). Hormonal influence on nuclear synthesis IV. Karyometric and microplotometric studies on rat thyroid nuclei in different functional states. *Acta Anat.* (*Basel*) **23**, 185–205.

Anders, F. (1967). Tumour formation in platyfish-swordtail hybrids as a problem of gene regulation. *Experientia* **23**, 1–10.

Anders, A., and Anders, F. (1963). Genetish bedingte XX- and XY- ♀ ♀ and XY- and YY- ♂ ♂ beim wilden *Platypoecilus maculatus* aus Mexico. *Z. Vererbungslehre* **94**, 1–18.

Anders, A., Anders, F., and Rase, S. (1969a). XY females caused by X-irradiation. *Experientia* **25**, 871.

Anders, A., Anders, F., Förster, W., Klinke, K., and Rase, S. (1969b). XX-, XY-, YY- ♀ ♀ and XX-, XY-, YY- ♂ ♂ bei *Platypoecilus maculatus* (Poeciliidae). *Z. Anz. Suppl.* **33**, 333–339.

Anderson, J. F. (1967). Histopathology of intersexuality in mosquitoes. *J. Exp. Zool.* **165**, 475–495.

Arrighi, F. E., and Hsu, T. C. (1971). Localization of heterochromatin in human chromosomes. *Cytogenetics* **10**, 81–86.

Ashburner, M. (1970). Patterns of puffing activity in the salivary gland chromosomes of *Drosophila*. V. Responses to environmental treatments. *Chromosoma* **31**, 356–376.

Astauroff, B. L. (1929). Studien über die erbliche Veränderung der Halteren bei *Drosophila melanogaster* Schin. *Arch. Entwicklungsmech. Organismen* **115**, 424–447.

Astauroff, B. L. (1930). Analyse der erblichen Störungsfälle der bilateralen Symmetrie im Zusammenhang mit der selbstständigen Variabilität ähnlicher Strukturen. *Z. Indukt. Abstamm. Vererbungsl.* **55**, 183–262.

Atkin, N. B., and Ohno, S. (1967). DNA values of four primitive chordates. *Chromosoma* **23**, 10–13.

Atkin, N. B., Mattinson, C., Beçak, W., and Ohno, S. (1965). The comparative DNA content of 19 species of placental mammals, reptiles and birds. *Chromosoma* **17**, 1–10.

Atz, J. W. (1964). Intersexuality in fishes. *In* "Intersexuality" (C. N. Armstrong and A. J. Marshall, eds.), pp. 145–232. Academic Press, New York.

Austin, C. R. (1961). "The Mammalian Egg." Blackwell Sci. Publ., Oxford.

Austin, C. R., and Amoroso, E. C. (1957). Sex chromatin in early cat embryos. *Exp. Cell Res.* **13**, 419–421.

Avery, A. G., Satina, S., and Rietsema, J. (1959). "Blakeslee: The Genus *Datura*." Ronald Press, New York.

Avery, O. T., MacLeod, C. M., and McCarty, M. (1944). Studies on the chemical nature of the substance inducing transformation of pneumococcal types. *J. Exp. Med.* **79**, 137–158.

Ayonoadu, U. W., and Rees, H. (1968). The regulation of mitosis by B-chromosomes in rye. *Exp. Cell Res.* **52**, 284–290.

Bacci, G. (1965). "Sex Determination." Pergamon, New York.

Baikie, A. G., Dartnall, J. A., and Lickiss, J. N. (1972). XXY son of a possible XX/XXY mother. *Lancet* i, 697–698.

Bain, A. D., and Scott, J. S. (1965). Mixed gonadal dysgenesis with XX/XY mosaicism. *Lancet* i, 1035–1039.

Baker, W. K. (1968). Position—effect variegation. *Advan. Genet.* **14**, 133–169.

Balkashina, E. I. (1929). Ein Fall der Erbhomöosis (die Genvariation "Aristopedia") bei *Drosophila melanogaster*. *Arch. Entwicklungsmech. Organismen* **115**, 448–463.

Baltzer, F. (1914). Die Bestimmung des Geschlechts nebst einer Analyse des Geschlechtsdimorphismus bei *Bonellia*. *Mitt. Zool. Stat. Neapel* **22**, 1–44.

Baltzer, F. (1937). Entwicklungsmechanische Untersuchungen an *Bonellia viridis* III. Über die Entwicklung und Bestimmung des Geschlechts und die Anwendbarkeit des Goldschmidtschen Zeitgesetzes der Intersexualität bei *Bonellia viridis*. *Pubbl. Sta. Zool. Napoli* **16**, 89–159.

Barigozzi, C., Dolfini, S., Fraccaro, M., Raimondi, R., and Tiepolo, L. (1966). *In vitro* studies of the DNA replication patterns of somatic chromosomes of *Drosophila melanogaster*. *Exp. Cell Res.* **43**, 231–234.

Barlow, P. W. (1972). Differential cell division in human X chromosome mosaics. *Humangenetik* **14**, 122–127.

Barr, H. J., and Ellison, J. R. (1971). Quinacrine staining of chromosomes and evolutionary studies in *Drosophila. Nature (London)* **233**, 190–191.

Barr, M. L., and Bertram, E. G. (1949). A morphological distinction between neurones of the male and female, and the behaviour of the nucleolar satellite during accelerated nucleoprotein synthesis. *Nature (London)* **163**, 676–677.

Barr, M. L., Shaver, E. L., and Carr, D. H. (1959). An unusual sex chromatin pattern in three mentally deficient subjects. *J. Ment. Defic. Res.* **4**, 89–107.

Barr, M. L., Carr, D. H., Pozony, J., Wilson, R. A., Dunn, H. G., Jacobson, T. S., and Miller, J. B. (1962). The XXXXY sex chromosome abnormality. *Can. Med. Ass. J.* **87**, 891–901.

Barr, M. L., Sergovich, F. R., Carr, D. H., and Shaver, E. L. (1969). The triplo-X female: an appraisal based on a study of 12 cases and a review of the literature. *Can. Med. Ass. J.* **101**, 247–258.

Bateson, W. (1894). "Materials for the study of Variation Treated with Especial Regard to Discontinuity." Macmillan, New York.

Bateson, W. (1909). "Mendel's Principles of Heredity." Cambridge University Press, London.

Bateson, W., and Saunders, E. R. (1902). Experimental studies in the physiology of heredity. *Rep. Evol. Comm. Roy. Soc.* **1**, 1–160.

Bateson, W., Saunders, E. R., and Punnett, R. C. (1908). Experimental studies in the physiology of heredity. *Rep. Evol. Comm. Roy. Soc.* **4**, 1–18.

Beadle, G. W., and Tatum, E. L. (1941). Genetic control of biochemical reactions in *Neurospora. Proc. Nat. Acad. Sci. U.S.* **27**, 499–506.

Beatty, R. A. (1957). "Parthenogenesis and Polyploidy in Mammalian Development." Cambridge University Press, London.

Beermann, W. (1967). Gene action at the level of the chromosome. *In* "Heritage from Mendel" (R. A. Brink, ed.), pp. 179–201. University of Wisconsin Press, Madison, Wisconsin.

Beermann, W., and Clever, U. (1964). Chromosome puffs. *Sci. Amer.* **210**, 50–58 (June).

Bellairs, R. (1971). "Developmental Processes in Higher Vertebrates." Logos Press, London.

Bellamy, A. W. (1922). Sex-linked inheritance in the teleost *Platypoecilus maculatus. Anat. Rec.* **24**, 419–420.

Benirschke, K. (1972). Chimerism, mosaicism and hybrids. *Proc. 4th Int. Congr. Human Genet. Paris*, Excepta Medica, Amsterdam.

Bennett, M. D. (1972). Nuclear DNA content and minimum generation time in herbivorous plants. *Proc. Roy. Soc. London* **B181**, 109–135.

Bennett, M. D. (1970). Natural variation in nuclear characters of meristems in *Vicia faba. Chromosoma* **29**, 317–335.

Bennett, M. D., and Rees, H. (1967). Natural and induced changes in chromosome size and mass in meristems. *Nature (London)* **215**, 93–94.

Bennett, M. D., and Rees, H. (1969). Induced and developmental variation in chromosomes of meristematic cells. *Chromosoma* **27**, 226–244.

Bernard, R., Stahl, A., Coignet, J., Giraud, F., Hartung, M., Brusquet, Y.,

and Passeron, P. 1969). Triploidie chromosomique chez un nouveau-né
polymalformé. *Ann. Genet.* **10,** 70.

Bierich, J. R. (1963). The adrenogenital syndrome. *In* "Intersexuality"
(C. Overzier, ed.), pp. 345–383. Academic Press, New York.

Bishop, D. W. (1961). Biology of spermatozoa. *In* "Sex and Internal Secretions" (W. C. Young, ed.), 3rd ed., Vol. 2, pp. 3–75. Bailliere, Tindall
& Cox, London.

Bloch, E. (1967). Sex hormones. *In* "The Encyclopedia of Biochemistry"
(R. J. Williams and E. M. Lansford, eds.), pp. 744–748. Reinhold,
New York.

Bloom, S. E. (1971). Private communication.

Boczkowski, K. (1971). Sex determination and gonadal differentiation in man.
*Clin. Genet.* **2,** 379–386.

Bodenstein, D., and Abdel-Malek, A. (1949). The induction of aristapedia
by nitrogen mustard in *Drosophila virilis. J. Exp. Zool.* **111,** 95–115.

Boivin, A., Vendrely, R., and Vendrely, C. (1948). L'acide désoxyribo-
nucléique du noyau dépositaire des caractères héréditaires; arguments
d'ordre analytique. *C. R. Acad. Sci. Paris* **236,** 1061–1063.

Bomsel-Helmreich, O. (1970). Fate of heteroploid embryos. *In* "Advances
in Biosciences," Vol. 6. Schering Symposium on Intrinsic and Extrinsic
Factors in Early Mammalian Development (G. Raspe, ed.), pp. 381–403
Pergamon, New York.

Boveri, T. (1891). Befruchtung. *Ergeb. Anat. Entwicklungmech.* **1** (Abt.
2), 386–485.

Breeuwsma, A. J. (1968). A case of XXY chromosome constitution in an
intersex pig. *J. Reprod. Fert.* **16,** 119–120.

Bridges, C. B. (1914). Direct proof through non-disjunction that the sex-linked
genes are borne on the X-chromosome. *Science* **40,** 107–109.

Bridges, C. B. (1916). Non-disjunction as proof of the chromosome theory
of heredity. *Genetics* **1,** 1–52 and 107–163.

Bridges, C. B. (1921a). Triploid intersexes in *Drosophila melanogaster. Science*
**54,** 252–254.

Bridges, C. B. (1921b). Genetical and cytological proof of nondisjunction
of the fourth chromosome of *Drosophila melanogaster. Proc. Natl. Acad.
Sci. U.S.* **7,** 186–192.

Bridges, C. B. (1922). The origin of variations in sexual and sex-limited
characters. *Amer. Natur.* **56,** 51–63.

Bridges, C. B. (1925). Sex in relation to chromosomes and genes. *Amer.
Natur.* **59,** 127–137.

Bridges, C. B. (1939). Cytological and genetic basis of sex. *In* "Sex and
Internal Secretions" (E. Allen, ed.), 2nd ed., pp. 15–63. Bailliere, London.

Bridges, C. B., and Dobzhansky, T. (1932). The mutant "proboscipedia"
in Drosophila melanogaster—a case of hereditary homöosis. *Arch. Entwicklungsmech. Organismen* **127,** 575–590.

Bridges, C. B., and Morgan, T. (1923). The third chromosome group of
mutant characters in *Drosophila melanogaster. Carnegie Inst. Wash. Publ.
No.* 327.

Britten, R. J., and Davidson, E. H. (1969). Gene regulation for higher cells—a theory. *Science* **165**, 349–357.

Britten, R. J., and Kohne, D. E. (1969a). Repetition of nucleotide sequences in chromosomal DNA. *In* "Handbook of Molecular Cytology" (A. Lima-de-Faria, ed.), pp. 21–36. North Holland, Amsterdam.

Britten, R. J., and Kohne, D. E. (1969b). Implications of repeated nucleotide sequences. *In* "Handbook of Molecular Cytology" (A. Lima-de-Faria, ed.), pp. 37–51. North Holland, Amsterdam.

Brøgger, A., and Aagenaes, Ö. (1965). The human Y-chromosome and the etiology of true hermaphroditism. With the report of a case with XX/XY sex chromosome mosaicism. *Hereditas* **53**, 231–246.

Brown, D. D., and Dawid, I. B. (1968). Specific gene amplification in oocytes. *Science* **160**, 272–280.

Brown, D. D., and Gurdon, J. B. (1964). Absence of ribosomal RNA synthesis in the anucleate mutant of *Xenopus laevis*. *Proc. Nat. Acad. Sci. U.S.* **51**, 139–146.

Brown, S. W. (1966). Heterochromatin. *Science* **151**, 417–425.

Bruner, J. A., and Witschi, E. (1946). Testosterone-induced modifications of sex development in female hamsters. *Amer. J. Anat.* **79**, 293–320.

Buckton, K. E., and Cunningham, C. (1971). Variations of the chromosome number in the red fox (*Vulpes vulpes*). *Chromosoma* **33**, 268–272.

Burns, J. A., and Gerstel, D. U. (1967). Flower color variegation and instability of a block of heterochromatin in *Nicotiana*. *Genetics* **57**, 155–167.

Burns, R. K. (1950). Sex transformation in the opossum: some new results and a retrospect. *Arch. Anat. Microsc. Morphol. Exp.* **39**, 467–483.

Burns, R. K. (1955). Urinogenital systems. *In* "Analysis of Development" (B. H. Willier, P. A. Weiss, and V. Hamburger, eds.), pp. 462–491. Saunders, Philadelphia, Pennsylvania.

Burns, R. K. (1961). Role of hormones in the differentiation of sex. *In* "Sex and Internal Secretions" (W. C. Young, ed.), 3rd ed., pp. 76–158. Baillière, Tindall & Cox, London.

Butler, L. J., Chantler, C., and Keith, C. G. (1969). A liveborn infant with complete triploidy (69, XXX). *J. Med. Genet.* **6**, 413–421.

Butler, N. R., and Alberman, E. D. (1969). "Perinatal Problems." Livingstone, Edinburgh.

Callan, H. G. (1963). The nature of lampbrush chromosomes. *Int. Rev. Cytol.* **15**, 1–34.

Carlson, E. A. (1966). "The Gene: A Critical History." Saunders, Philadelphia, Pennsylvania.

Carr, D. H. (1971a). Chromosome studies in selected spontaneous abortions. Polyploidy in man. *J. Med. Genet.* **8**, 164–174.

Carr, D. H. (1971b). Genetic basis of abortion. *Ann. Rev. Genet.* **5**, 65–80.

Carter, C. O., David, P. A., and Laurence, K. M. (1968). A family study of major central nervous system malformations in South Wales. *J. Med. Genet.* **5**, 81–106.

Casey, M. D., Blank, C. E., Mobley, T., Kohn, P., Street, D. R. K.,

McDougall, J. M., Gooder, J., and Platts, J. (1971). Special Hospitals Research Report No. 2. Patients with chromosome abnormality in two special hospitals. Broadmore Hospital, Crowthorne, Berkshire.

Caspersson, T., Zech, L., Modest, E. J., Foley, G. E., Wagh, U., and Simonsson, E. (1969a). Chemical differentiation with fluorescent alkylating agents in *Vicia faba* metaphase chromosomes. *Exp. Cell Res.* **58**, 128–140.

Caspersson, T., Zech, L., Modest, E. J., Foley, G. E., Wagh, U., and Simonsson, E. (1969b). DNA-binding fluoro-chromes for the study of the organization of the metaphase nucleus. *Exp. Cell Res.* **58**, 141–152.

Castle, W. E. (1903). The laws of heredity of Galton and Mendel, and some laws governing race improvement by selection. *Proc. Amer. Acad. Arts Sci.* **39**, 223–242.

Catcheside, D. G. (1947). The P-locus position effect in *Oenothera*. *J. Genet.* **48**, 31–42.

Cattanach, B. M. (1961). XXY mice. *Genet. Res.* **2**, 156–158.

Cattanach, B. M. (1962). XO mice. *Genet. Res.* **3**, 487–490.

Cattanach, B. M., Pollard, C. E., and Hawkes, S. G. (1971). Sex reversed mice: XX and XO males. *Cytogenetics* **10**, 318–337.

Chambon, A. (1972). "La Triploidie chez l'Enfant. Essai d'Explication du Pseudo-Hermaphrodisme Mâle. A propos d'une Observation de Triploidie en Mosaique 46,XX/69,XXY." Lyon (unpublished).

Chang, C. Y. (1953). Parabiosis and gonad transplantation in *Xenopus laevis* Daudin. *J. Exp. Zool.* **123**, 1–27.

Chang, C. Y., and Witschi, E. (1956). Genetic control and hormonal reversal of sex differentiation in *Xenopus*. *Proc. Soc. Exp. Biol. Med.* **93**, 140–144.

Chargaff, E. (1950). Chemical specificity of nucleic acids and mechanisms of their enzymatic degradation. *Experientia* **6**, 201–209.

Chavin, W., and Gordon, M. (1951). Sex determination in *Platypoecilus maculatus* I. *Zoologica* **36**, 135–145.

Chen, T. Y. (1929). On the development of imaginal buds in normal and mutant *Drosophila melanogaster*. *J. Morphol.* **47**, 135–199.

Clough, E., Pyle, R. L., Hare, W. C. D., Kelly, D. F., and Patterson, D. E. (1970). An XXY sex-chromosome constitution in a dog with testicular hypoplasia and congenital heart disease. *Cytogenetics* **9**, 71–77.

Comings, D. E. (1971). Heterochromatin of the Indian Muntjac. *Exp. Cell Res.* **67**, 441–460.

Cooper, D. W., Vande Berg, J. L., Sharman, G. B., and Poole, W. E. (1971). Phosphoglycerate kinase polymorphism in kangaroos provides further evidence for paternal X inactivation. *Nature New Biol.* **230**, 155–157.

Cooper, K. W. (1959). Cytogenetic analysis of major heterochromatic elements (especially Xh and Y) in *Drosophila melanogaster*. *Chromosoma* **10**, 535–588.

Corner, G. W. (1947). "The Hormones in Human Reproduction." Princeton Univ. Press, Princeton, New Jersey.

Correns, C. (1900). G. Mendel's Regel über das Verhalten der Nachkommenschaft der Rassenbastarde. *Ber. Deut. Bot. Ges.* **18**, 146–168.

Correns, C. (1907). "Die Bestimmung und Vererbung des Geschlechts, nach neuen Versuchen mit höheren Pflanzen." Borntraeger, Berlin.

Correns, C. (1913). "Die Vererbung und Bestimmung des Geschlechtes." Borntraeger, Berlin.

Correns, C. (1928). *In* "Bestimmung Vererbung and Verteilung des Geschlechtes bei den höheren Pflanzen," Vol. 2: Handbuch der Vererbungswissenschaft (E. Baur and M. Hartmann, eds.). Borntraeger, Berlin.

Courot, M., Hochereau-de Reviers, M. T., and Ontavant, R. (1970). Spermatognesis. *In* "The Testis" (A. D. Johnson, W. R. Gomes, and N. L. Vandemark, eds.), Vol. 1, pp. 339–432. Academic Press, New York.

Court Brown, W. M. (1968). Males with an XYY sex chromosome constitution. *J. Med. Genet.* **5**, 341–359.

Court Brown, W. M., Jacobs, P. A., and Brunton, M. (1965). Chromosome studies on randomly chosen men and women. *Lancet* **ii**, 561–562.

Crick, F. H. C. (1966). The genetic code: III. *Sci. Amer.* **215**, 55–74 (June).

Crick, F. (1971). General model for the chromosomes of higher organisms. *Nature* (*London*) **234**, 25–27.

Cuénot, L. (1899). Sur la détermination due sexe chez les animaux. *Bull. Sci. Fr. Belg.* **32**, 461–534.

d'Ancona, U. (1950). Détermination e differenciation du sexe chez les poissons. *Arch. Anat. Microsc. Morphol. Exp.* **39**, 274–294.

Dantschakoff, V. (1941). "Der Aufbau des Geschlechts bei höheren Wirbeltieren." Fischer, Jena.

Darlington, C. D. (1958). "Evolution of Genetic Systems," 2nd ed. Oliver and Boyd, Edinburgh.

Darlington, C. D., and Haque, A. (1966). Organisation of DNA synthesis in chromosomes. *In* "Chromosomes Today" (C. D. Darlington and K. R. Lewis, eds.), pp. 102–107. Oliver and Boyd, Edinburgh.

Darlington, C. D., and Thomas, P. T. (1941). Morbid mitosis and the activity of inert chromosomes in *Sorghum. Proc. Roy. Soc.* **B130**, 127–150.

Davidson, E. H., Crippa, M., Kramer, F. R., and Mirsky, A. E. (1966). Genomic Function during the lampbrush chromosome stage of amphibian oogenesis. *Proc. Nat. Acad. Sci. U.S.* **56**, 856–863.

Davidson, R. G., Nitowsky, H. M., and Childs, B. (1963). Demonstration of two populations of cells in the human female heterozygous for glucose-6-phosphate dehydrogenase variants. *Proc. Nat. Acad. Sci. U.S.* **50**, 481–485.

de la Chapelle, A. (1972). Nature and origin of males with XX sex chromosomes. *Amer. J. Human Genet.* **24**, 71–105.

de Man, J. C. H., and Noorduyn, N. J. A. (1969). Ribosomes: properties and function. *In* "Handbook of Molecular Cytology (A. Lima-de Faria, ed.), pp. 1079–1100. North Holland, Amsterdam.

Denver Conference (1960). A proposed standard system of nomenclature of human chromosomes. *Lancet* **i**, 1063–1065.

de Vries, H. (1889). "Intracelluläre Pangenesis." Fisher, Jena.

de Vries, H. (1900). Das Spaltungsgesetz der Bastarde. *Ber. Deut. Bot. Ges.* **18**, 83–90.

Deys, B. B., Grzeschick, K. H., Grzeschick, A., Jaffé, E. R., and Siniscalco, M. (1972). Human phosphoglycerate kinase and inactivation of the X chromosome. *Science* **175**, 1002–1003.

Diaso, R. B., and Glass, R. H. (1970). The Y chromosome in sperm of an XYY male. *Lancet* **ii**, 1318–1319.

Diczfalusy, E., and Mancuso, S. (1969). Oestrogen metabolism in pregnancy. In "Foetus and Placenta" (A. Klopper and E. Diczfalusy, eds.), pp. 191–248. Blackwell, Oxford.

Dobzhansky, T. (1932). Cytological map of the X-chromosome of *Drosophila melanogaster*. *Biol. Zentralbl.* **52**, 493–509.

Dobzhansky, T. (1944). Distribution of heterochromatin in the chromosomes of *Drosophila pallidipennis*. *Amer. Natur.* **78**, 193–213.

Dobzhansky, T., and Schultz, J. (1934). The distribution of sex-factors in the X-chromosome of *Drosophila melanogaster*. *J. Genet.* **28**, 349–386.

Doncaster, L. (1914). "The Determination of Sex." Cambridge Univ. Press, London.

Doncaster, L., and Raynor, G. H. (1906). On breeding experiments with Lepidoptera. *Proc. Zool. Soc. London*, pp. 125–133.

Doniach, L., and Pelc, S. R. (1950). Autoradiograph technique. *Brit. J. Radiol.* **23**, 184–192.

Drets, M. E., and Shaw, M. W. (1971). Specific banding patterns of human chromosomes. *Proc. Nat. Acad. Sci. U.S.* **68**, 2073–2077.

Dunn, L. C. (1965). "A Short History of Genetics: The Development of some of the Main Lines of Thoughts, 1864–1939." McGraw-Hill, New York.

Düsing C. (1884). Die Regulierung des Geschlechtsverhältnisses bei der Vermehrung der Menschen, Tiere und Pflanzen. *Jena Z. Naturwiss.* **17**, 593–940.

Dzwillo, M. (1959). Quoted by Haskins et al. (1970).

East, L. M. (1910). A Mendelian interpretation of variation that is apparently continuous. *Amer. Natur.* **44**, 65–82.

Edwards, J. H. (1960). The simulation of Mendelism. *Acta Genet.* (*Basel*) **10**, 63–70.

Edwards, J. H. (1970). The operation of selection. In "Human Population Cytogenetics" (P. A. Jacobs, W. H. Price, and P. Law, eds.), Pfizer Med. Monograph 5, pp. 241–262. Edinburgh Univ. Press, Edinburgh.

Edwards, J. H., Yuncken, C., Rushton, D. I., Richards, S., and Mittwoch, U. (1967). Three cases of triploidy in man. *Cytogenetics* **6**, 81–104.

Ellison, J. R., and Barr, H. J. (1972). Quinacrine fluorescence of specific chromosome regions. Late replication and high A:T content in *Samoaia leonensis*. *Chromosoma* **36**, 375–390.

Elsdale, T. R., Fischberg, M., and Smith, S. (1958). A mutation that reduces nucleolar number in *Xenopus laevis*. *Exp. Cell Res.* **14**, 642.

Emerson, R. A., and East, E. M. (1913). The inheritance of quantitative characters in maize. *Res. Bull. Neb. Agr. Exp. Sta.* **2**, 1–120.

Epifanova, O. I. (1971). Effects of hormones on the cell cycle. In "The Cell Cycle and Cancer" (R. Baserga, ed.), pp. 144–190. Marcel Dekker, New York.

Evans, E. P., Ford, C. E., Chaganti, R. S. K., Blank, C. E., and Hunter, H. (1970). XY spermatocytes in an XYY male. *Lancet* i, 718–720.

Evans, H. J., and Sumner, A. T. (1973). Chromosome architecture; morphological and molecular aspects of longitudinal differentiation. *In* "Chromosomes Today" (J. Wahrman, ed.), Vol. 4. Oliver and Boyd, Edinburgh and London. In press.

Evans, H. J., Buckton, K. E., and Sumner, A. T. (1971). Cytological mapping of human chromosomes: results obtained with quinacrine fluorescence and the acetic-saline-giemsa technique. *Chromosoma* 35, 310–325.

Fahmy, O. G., and Fahmy, M. J. (1966). The nature and distribution of the mutations induced by unirradiated and irradiated heterologous deoxyribonucleic acid in *Drosophila melanogaster*. *Genetics* 54, 1123–1138.

Falconer, D. S. (1964). "Introduction to Quantitative Genetics." Oliver and Boyd, Edinburgh.

Farnsworth, M. W. (1965). Growth and cytochrome c oxidase activity in Minute nutants of *Drosophila*. *J. Exp. Zool.* 160, 355–361.

Farnsworth, M. W., and Jozwiak, J. (1969). Terminal respiration in *Minute* mutants of *Drosophila*. *J. Exp. Zool.* 171, 119–126.

Fankhauser, C. (1945). The effects of changes in chromosome number on amphibian development. *Quart. Rev. Biol.* 20, 20–78.

Ferrier, P., Ferrier, S., Stalder, G. Bühler, E., Bamatter, F., and Klein, D. (1964). Congenital asymmetry associated with diploid/triploid mosaicism, and large satellites. *Lancet* 1, 80–82.

Fedrick, J., Alberman, E. D., and Goldstein, H. (1971). Possible teratogenic effect of cigarette smoking. *Nature (London)* 231, 529–530.

Ferguson-Smith, M. A. (1966). X-Y chromosome interchange in the aetiology of true hermaphroditisus and of XX Klinefelter's syndrome. *Lancet* ii, 475–476.

Fincham, J. R. S. (1970). The regulation of gene mutation in plants. *Proc. Roy. Soc.* B176, 295–302.

Fischberg, M., and Beatty, R. A. (1951). Spontaneous heteroploidy in mouse embryos up to mid term. *J. Exp. Zool.* 118, 321–335.

Fishelson, L. (1970). Protogynous sex reversal in the fish *Anthias squamipinnis* (Teleostei, Anthiidae) regulated by the presence or absence of a male fish. *Nature (London)* 227, 90–91.

Fisher, R. A. (1918). The correlation between relatives on the supposition of Mendelian inheritance. *Trans. Roy. Soc. Edinburgh* 52, 399–433.

Fitzgerald, P. H., Brehaut, L. A., Shannon, F. T., and Angus, H. B. (1970). Evidence of XX/XY sex chromosome mosaicism in a child with true hermaphroditism. *J. Med. Genet.* 7, 383–388.

Flamm, W. G., McCallum, M., and Walker, P. M. B. (1967). The isolation of complimentary strands from a mouse DNA fraction. *Proc. Nat. Acad. Sci. U.S.* 57, 1729–1734.

Foote, C. L. (1964). Intersexuality in Amphibians. *In* "Intersexuality in Vertebrates including Man" (C. N. Armstrong and A. J. Marshall, eds.), pp. 233–272. Academic Press, New York.

Forbes, T. R. (1961). Endocrinology of reproduction in cold-blooded vertebrates. *In* "Sex and Internal Secretions" (W. C. Young, ed.), 3rd ed.,
pp. 1035–1087. Baillière, London.

Ford, C. E. (1969). Mosaics and chimaeras. *Brit. Med. Bull.* **25**, 104–109.

Ford, C. E. (1970a). The cytogenetics of the male germ cells and the testis
in mammals. *In* "The Human Testis" (E. Rosenberger and C. A. Paulsen,
eds.), pp. 139–149. Plenum, New York.

Ford, C. E. (1970b). Cytogenetics of sex determination in man and mammals.
*J. Biosoc. Sci. Suppl.* **2**, 7–30.

Ford, C. E., and Hamerton, J. L. (1956). The chromosomes of man. *Nature*
(*London*) **178**, 1020–1023.

Ford, C. E., Jones, K. W., Polani, P. E., Almeida, J. C., and Biggs, J. H.
(1959). A sex-chromosome anomaly in a case of gonadal dysgenesis.
(Turner's syndrome). *Lancet* **i**, 711–713.

Forteza, G., Bonilla, F., Baguena, R., Monmeneu, S., Galbis, M., and
Zaragoza, V. (1963). Un caso de mosaicismo XY/XX cromatinnegativo,
con disgenesia gonadal y sexo fenotypico feminino. *Rev. Clin. Espan.* **88**,
394–398.

Fraccaro, M., and Lindsten, J. (1960). A child with 49 chromosomes. *Lancet*
**ii**, 1303.

Fraccaro, M., Tiepolo, L., Zuffardi, O., Barigozzi, C., and Dolfini, S., (1971).
Fluorescence and Y translocation in XX male. *Lancet* **i**, 858.

Fraser, F. C., and Pashayan, H. (1970). Relation of face shape to susceptibility
to congenital cleft lip. *J. Med. Genet.* **7**, 112–117.

Fréderic, J. (1-961). Contribution a l'étude du caryotype chez le poulet.
*Arch. Biol.* **72**, 185–209.

Fredga, K. (1970). Unusual sex chromosome inheritance in mammals. *Phil.
Trans. Roy. Soc. London* **259**, 15–36.

Frizzi, G. (1948). L'eteropicnosi come indice di riconiscimento dei sessi in
"Bombyx mori L." *Ric. Sci.* **18**, 119–123.

Galton, F. (1889). "Natural Inheritance." Macmillan, New York.

Ganner, E., and Evans, H. J. (1971). The relationship between patterns
of DNA replication and of quinacrine fluorescence in the human chromosome complement. *Chromosoma* **35**, 326–341.

Garrod, A. E. (1908). The Croonian Lectures on Inborn Errors of Metabolism.
*Lancet* **ii**, 1–7, 73–79, 142–148, 214–220. Reprinted in Harris, H., "Garrod's
Inborn Errors of Metabolism," Oxford University Press, London (1963).

Gartler, S. M., Waxman, S. H., and Giblett, E. (1962). An XX/XY human
hermaphrodite resulting from double fertilization. *Proc. Nat. Acad. Sci.
U.S.* **48**, 332–335.

Geisler, M., Svejcar, J., and Degenhandt, K. H. (1972). XXY son of a XX/XXX
mother. *Lancet* **i**, 955.

George, D. P., and Polani, P. E. (1970). Y heterochromatin and XX males.
*Nature* (*London*) **228**, 1215–1216.

George, K. P. (1970). Cytochemical differentiation along human chromosomes.

*Nature* (*London*) **226**, 80–81.

German, J. L. (1962). DNA synthesis in human chromosomes. *Trans. N.Y. Acad. Sci.* **24** (2), 395–407.

German, J. L. (1967). Autoradiographic studies of human chromosomes I. A review. *Proc. 3rd Int. Cong. Human Genet., Chicago, 1966,* pp. 123–136.

German, J. (1970). Abnormalities of human sex chromosomes V. A unifying concept in relation to the gonadal dysgeneses. *Clin. Genet.* **1**, 15–27.

German, J., and Simpson, J. L. (1971). Abnormalities of human autosomes I Ambiguous genitalia associated with a translocation 46, XY, t (Cq+; Cq-). *Birth Defects Orig. Artic. Ser.* **7**, 145–149.

Gianelli, F. (1963). The pattern of X-chromosome deoxyribonucleic acid synthesis in two women with abnormal sex chromosome complements. *Lancet* i, 863–865.

Gier, H. T., and Marion, G. B. (1970). Development of the mammalian testis. *In* "The Testis" (A. D. Johnson, W. B. Gomes, and N. L. Vandemark, eds.), pp. 1–45. Academic Press, New York.

Glenister, T. W. (1956). Determination of sex in early human embryos. *Nature* (*London*) **177**, 1135–1136.

Goldfeder, A. (1965). Biological properties and radiosensitivity of tumours: determination of the cell-cycle and time of synthesis of deoxyribonucleic acid using tritiated thymidine and autoradiography. *Nature* (*London*) **207**, 612–614.

Goldschmidt, R. B. (1920). "Die quantitative Grundlage von Vererbung und Artbildung." Vorträge und Aufsätze über Entwicklungsmechanik der Organismen (W. Roux, ed.). Springer-Verlag, Berlin.

Goldschmidt, R. B. (1934). *Lymantria. Bibl. Genet.* (*Leipzig*) **11**, 1–180.

Goldschmidt, R. B. (1935). Gen und Ausseneigenschaft. I and II. *Z. Indukt. Abstamm. Vererbungsl.* **69**, 38–69 and 70–131.

Goldschmidt, R. B. (1940). "The Material Basis of Evolution." Yale Univ. Press, New Haven, Connecticut.

Goldschmidt, R. B. (1949). Heterochromatic heredity. *Proc. 8th Int. Congr. Genet. Stockholm, 1948,* pp. 244–255.

Goldschmidt, R. B. (1954). Different philosophies of genetics. *Proc. 9th Int. Congr. Genet.* (Caryologia 6, Suppl.), Italy, Vol. 1. pp. 83–99.

Goldschmidt, R. B. (1955). "Theoretical Genetics." University of California Press, Berkeley and Los Angeles, California.

Goldschmidt, R. B., and Piternick, L. K. (1957a). The genetic background of chemically induced phenocopies in *Drosophila. J. Exp. Zool.* **135**, 127–202.

Goldschmidt, R. B., and Piternick, L. K. (1957b). The genetic background of chemically induced phenocopies in *Drosophila.* II. *J. Exp. Zool.* **136**, 201–228.

Goldschmidt, R. B., Hannah, A., and Piternick, L. K. (1951). The podoptera effect in *Drosophila melanogaster. Univ. Calif. Press. Publ. Zool.* **55**, 67–293.

Goodman, R. M., Miller, F., and North, C. (1968). Chromosomes of tall men. *Lancet* i, 1318.

Gordon, M. (1927). The genetics of a viviparous top minnow, *Platypoecilus;* the inheritance of two kinds of melanophores. *Genetics* 12, 253–283.

Gordon, M. (1937). Genetics of *Platypoecilus*. III. Inheritance of sex and crossing over of the sex chromosomes in the platyfish. *Genetics* 22, 376–392.

Gordon, M. (1947). Genetics of *Platypoecilus maculatus*. IV. The sex determining mechanism in two wild populations of the Mexican platyfish. *Genetics* 32, 8–17.

Gordon, M. (1948). Effects of five primary genes on the site of melanomas in fishes and the influence of two color genes on their pigmentation. *Spec. Publ. N.Y. Acad. Sci.* 4, 216–268.

Gordon, M. (1950). Fishes as laboratory animals. *In* "The Care and Breeding of Laboratory Animals" (E. Farris, ed.), pp. 345–449. Wiley, New York.

Gordon, M. (1952). Sex determination in *Xiphophorus* (*Platypoecilus*) *maculatus*. III. Differentiation of gonads in platyfish from broods having a sex ratio of three females to one male. *Zoologica* 37, 91–100.

Gordon, M. (1959). The melanoma cell as an incompletely differentiated pigment cell. *In* "Pigment Cell Biology" (M. Gordon, ed.), pp. 215–239. Academic Press, New York.

Grobstein, C. (1948). Optimal gonopodial morphogenesis in *Platypoecilus maculatus* with constant dosage of methyl testosterone. *J. Exp. Zool.* 109, 215–233.

Gropp, H., Hein, B., and Wolf, U. (1966). Kerngrösse und Sex-Chromatinhäufigkeit beim Mammacarcinom. *Z. Krebsforsch.* 68, 123–130.

Gross, J. (1906). Die Spermatogenese von *Pyrrhocoris. Zool. Jahrb. Anat. Abt. 23.*

Grumbach, M. M., Morishima, A., and Taylor, J. H. (1963). Human sex chromosome abnormalities in relation to DNA replication and heterochromatinization. *Proc. Nat. Acad. Sci. U.S.* 49, 581–589.

Grüneberg, H. (1938). An analysis of the "pleiotropic" effects of a new lethal mutation in the rat (*Mus norvegicus*). *Proc. Roy. Soc.* B125, 123–144.

Grüneberg, H. (1943). Congenital hydrocephalus in the mouse (a case of spurious pleiotropism). *J. Genet.* 45, 1–21.

Grüneberg, H. (1952). Genetical studies on the skeleton of the mouse. IV Quasi-continuous variation. *J. Genet.* 51, 95–114.

Grüneberg, H. (1955). Genetical studies on the skeleton of the mouse. XV Relations between major and minor variants. *J. Genet.* 13, 515–535.

Grüneberg, H. (1964). The genesis of skeletal abnormalities. *In* "Congenital Malformations" (M. Fishbein, ed.), pp. 219–223. Int. Med. Cong., New York.

Gutherz, S. (1970). Zur Kenntnis der Heterochromosomen. *Arch. Mikroskop. Anat. Entwicklungsmech.* 69, 491–514.

Hadorn, E. (1968). Transdetermination in cells. *Sci. Amer.* 219, 110–120.

Hadlane, J. B. S. (1927). "Possible Worlds," pp. 18–26. Chatto & Windus, London.

Haldane, J. B. S. (1932). "The Causes of Evolution." Allen & Unwin, London.

Hall, B. (1965). Delayed ontogenesis in human trisomy syndromes. *Hereditas* **52**, 334–344.

Hall, B. (1966). Follow-up investigation of newborn mongoloids with respect to growth retardation. *Hereditas* **56**, 99–108.

Hamerton, J. L. (1968). The significance of sex chromosome derived heterochromatin in mammals. *Nature (London)* **219**, 910–914.

Hamerton, J. L. (1971a). "Human Cytogenetics," Vol. 1. Academic Press, New York.

Hamerton, J. H. (1971b). "Human Cytogenetics," Vol. 2, Clinical Cytogenetics. Academic Press, New York.

Hamerton, J. L., and Gianelli, F. (1970). Non-random inactivation of the X chromosome in the female mule. *Nature* **228**, 1322–1323.

Hamerton, J. L., Gianelli, F., Collins, F., Hallett, J., Fryer, A., McGuire, V. M., and Short, R. V. (1969a). Non-random X-inactivation in the female mule. *Nature (London)* **222**, 1277–1278.

Hamerton, J. L., Dickson, J. M., Pollard, C. E., Grieves, S. A., and Short, R. V. (1969b). Genetic intersexuality in goats. *J. Reprod. Fert. Suppl.* **7**, 25–51.

Hamerton, J. L., Richardson, B. J., Gee, P. A., Allen, W. R., and Short, R. V. (1971). Non-random X chromosome expression in female mules and hinnies. *Nature* **232**, 312–315.

Handmaker, S. D. (1963). The satellited chromosomes of man with reference to the Marfan syndrome. *Amer. J. Human Genet.* **15**, 11–18.

Harnden, D. G. (1961). Nuclear sex in triploid XXY human cells. *Lancet* **ii**, 488.

Harrington, R. W., Jr. (1961). Oviparous hermaphroditic fish with internal self-fertilization. *Science* **134**, 1749–1750.

Harrington, R. W., Jr. (1968). Delimination of the thermolabile phenocritical period of sex determination and differentiation in the outogeny of the normally hermaphroditic fish *Rivulus marmoratus* Poey. *Physiol. Zool.* **41**, 447–460.

Harrington, R. W., Jr. (1971). How ecological and genetic factors interact to determine when self-fertilised hermaphrodites of *Rivulus marmoratus* change into functional secondary males, with a re-appraisal of the modes of intersexuality among fishes. *Coppeia* **3**, 389–432.

Harrington, R. W., Jr., and Kallman, K. D. (1968). The homozygosity of clones of the self-fertilizing hermaphroditic fish *Rivulus marmoratus* Poey (Cypronodontidae, Atheriniformes). *Amer. Natur.* **102**, 337–343.

Harris, H. (1970). "The Principles of Human Biochemical Genetics." North Holland, Amsterdam.

Hartmann, M. (1956). "Die Sexualität," 2nd ed. Fischer, Stuttgart.

Hartmann, M., and Huth, W. 1936). Untersuchungen über Geschlechtsbestimmung und Geschlechtsumwandlung von *Ophryotrocha puerilis. Zool. Jahrb.* (Abt. Zool. Physiol.) **56**, 389–439.

Haskins, C. P., Young, P., Hewitt, R. E., and Haskins, E. F. (1970). Stabilised heterozygosis of supergenes mediating certain Y-linked colour patterns in populations of *Lebistes reticulatus. Heredity* **25**, 575–589.

Hauser, G. A. (1963a). Gonadal dysgenesis. *In* "Intersexuality," (C. Overzier, ed.), pp. 298–339. Academic Press, London and New York.

Hauser, G. A. (1963b). Testicular feminization. *In* "Intersexuality," (C. Overzier, ed.), pp. 255–276. Academic Press, London and New York.

Hayes, W. (1952). Recombination in *Bact. coli* K 12: unidirectional transfer of genetic material. *Nature (London)* **169**, 118–119.

Hayes, W. (1968). "The Genetics of Bacteria and their Viruses," 2nd ed. Blackwell Sci. Publ., Oxford.

Heitz, E. (1928). Das Heterochromatin der Moose, I. *Jahrb. Wiss. Bot.* **69**, 762–818.

Heitz, E. (1931). Die Ursache der gesetzmässigen Zahl, Lage, Form und Grösse der pflanzlichen Nukleolen. *Planta* **12**, 774–844.

Heitz, E. (1932). Nukleolen und Chromosomen in der Gattung *Vicia. Planta* **12**, 774–844.

Heitz, E. (1933). Die somatische Heteropyknose bei *Drosophila melanogaster* und ihre genetische Bedeutung. *Z. Zellforsch. Mikrosk. Anat.* **20**, 237–287.

Heitz, E. (1934). Uber α- und β-Heterochromatin sowie Konstanz und Bau der Chromomeren bei *Drosophila. Biol. Zentrabl.* **54**, 588–609.

Heitz, E. (1935). Chromosomenstruktur und Gene. *Z. Indukt. Abstamm. Vererbungsl.* **70**, 402–447.

Henking, H. (1891). Untersuchungen über die ersten Entwicklungsvorgänge in den Eiern der Insekten. II. Über Spermatogenese und deren Beziehung zur Entwicklung bei *Pyrrhocoris apterus. Z. Wiss. Zool. Abt.* **A51**, 685–736.

Henneberg, B. (1897). Wodurch wird das Geschlechtsverhältniss beim Menschen und den höheren Tieren beeinflusst? *Ergeb. Anat. Entwicklungsmech.* **7**, 697–721.

Herbst, C. (1935). Untersuchungen zur Bestimmung des Geschlechts IV. Die Abhängigkeit vom Kaliumgehalt des umgebenden Mediums bei Bonellia viridis. *Arch. Entwicklungsmech. Organismen* **132**, 576–599.

Hess, O. (1970). Lampenbürstenchromosomen. *In* "Handbuch der Allgemeinen Pathologie" (H. W. Altmann *et al.*, eds.), Vol. 2, Part 2, pp. 215–281. Springer Verlag, Berlin.

Hess, O., and Meyer, G. F. (1968). Genetic activities of the Y chromosome in *Drosophila* during spermatogenesis. *Advan. Genet.* **14**, 171–223.

Hill, R. N., and Yunis, J. J. (1961). Mammalian X-chromosomes: change in pattern of DNA replication during embryogenesis. *Science* **155**, 1120–1121.

Hoffenberg, R., and Jackson, W. P. U. (1957). Gonadal dysgenesis: Modern concepts. *Brit. Med. J.* **2**, 1457–1462.

Holt, S. B. (1968). "The Genetics of Dermal Ridges." Thomas, Springfield, Illinois.

Holtzer, H., Bischoff, R., and Chacko, S. (1969). Activity of the cell surface during myogenesis and chondrogenesis. In "Cellular Recognition" (R. Smith and R. Good, eds.), pp. 19–25. North Holland, Amsterdam.

Hook, B., and Brustman, L. D. (1971). Evidence for selective differences between cells with an active horse X chromosome and cells with an active donkey X chromosome in the female mule. Nature (London) 232, 349–350.

Howard, A., and Pelc, S. R. (1953). Synthesis of deoxyribonucleic acid in normal and irradiated cells and its relation to chromosome breakage. Heredity Suppl. 6, 261–274.

Hsia, D. Y-Y., Inouye, T., Wong, P., and South, A. (1964). Studies on galactose oxidation in Down's syndrome. N. Engl. J. Med. 270, 1085–1088.

Hsu, T. C. (1962). Differential rate in RNA synthesis between euchromatin and heterochromatin. Exp. Cell. Res. 27, 332–334.

Hsu, T. C., and Benirschke, K. (1967). "An Atlas of Mammalian Chromosomes." Springer Verlag, Berlin.

Hughes, W. (1929). The freemartin condition in swine. Anat. Rec. 41, 213–245.

Hughes-Schrader, S. (1948). Cytology of coccids (Coccoidea, Homoptera). Advan. Genet. 2, 127–203.

Hughesdon, P. E., and Kumarasamy, T. (1970). Mixed germ cell tumours (gonadoblastomas) in normal and dysgenetic gonads. Virchows Arch. Path. Anat. A349, 258–280.

Hultén, M. (1970). Meiosis in XYY men. Lancet i, 717–718.

Hutt, F. B. (1949). "Genetics of the Fowl." McGraw-Hill, New York.

Ilbery, P. L. T., and Williams, D. (1967). Evidence of the freemartin condition in the goat. Cytogenetics 6, 276–285.

Ingram, V. M. (1957). Gene mutations in human haemoglobin: the chemical difference between normal and sickle cell haemoglobin. Nature (London) 180, 326–328.

Issa, M., Blank, C. E., and Atherton, G. W. (1969). The temporal appearance of sex chromatin and of the late-replicating X chromosome in the domestic rabbit. Cytogenetics 8, 219–237.

Izawa, M., Allfrey, V. G., and Mirsky, A. E. (1963). The relationship between RNA synthesis and loop structure in lampbrush chromosomes. Proc. Nat. Acad. Sci. U.S. 49, 544–551.

Jacob, F., and Wollman, E. L. (1961). "Sexuality and the Genetics of Bacteria." Academic Press, New York.

Jacobs, P. A., and Ross, A. (1966). Structural abnormalities of the Y chromosome in man. Nature (London) 210, 352–354.

Jacobs, P. A., and Strong (1959). A case of human intersexuality having a possible XXY sex-determining mechanism. Nature (London) 183, 302–303.

Jacobs, P. A., Baikie, A. G., Court Brown, W. M., MacGregor, T. N., Maclean, N., and Harnden, D. G. (1959a). Evidence for the existence of the human "superfemale." *Lancet* **ii**, 423–425.

Jacobs, P. A., Baikie, A. G., Court Brown, W. M., Forrest, H., Roy, J. R., Stewart, J. S. S., and Lennox, B. (1959b). Chromosomal sex in the syndrome of testicular feminisation. *Lancet* **ii**, 591–592.

Jacobs, P. A., Brunton, M., Court Brown, W. M., Doll, R., and Goldstein, H. (1963). Change of human chromosome count distributions with age: evidence for a sex difference. *Nature (London)* **197**, 1080–1081.

Jacobs, P. A., Brenton, M., Melville, M. M., Brittain, R. P., and McClemont, W. F. (1965). Aggressive behaviour, mental subnormality and the XYY male. *Nature (London)* **208**, 1351–1353.

Jacobs, P. A., Price, W. H., Court Brown, W. M., Brittain, R. P., and Whatmore, P. B. (1968). Chromosome studies on men in a maximum security hospital. *Ann. Human Genet.* **31**, 339–347.

Janke, H. (1888). "Die willkürliche Hervorbringung des Geschlechts bei Mensch und Hausthieren," 2nd ed. Zimmer Stuttgart.

Janssens, F. A. (1909). Spermatogénèse dans les Batraciens. V. La théorie de la chiasmatypie. Nouvelles interpretations des cinèses de maturations. *Cellule* **25**, 387–411.

Johannsen, W. (1909). "Elemente der Exakten Erblichkeitslehre." Fischer, Jena.

Jones, R. N., and Rees, H. (1969). An anomalous variation due to B chromosomes in rye. *Heredity* **24**, 265–271.

Jones, T. C. (1969). Anomalies of sex chromosomes in tortoiseshell male cats. *In* "Comparative Mammalian Cytogenetics" (K. Benirschke, ed.), pp. 414–433. Springer Verlag, New York.

Jost, A. (1947). Recherches sur la différenciation sexuelle de l'embryon de lapin. III. Role des gonades foetales dans la différenciation sexuelle somatique. *Arch. Anat. Morphol. Exp.* **36**, 271–315.

Jost, A. (1965). Gonadal hormones in the sex differentiation of the mammalian fetus. *In* "Organogenesis" (R. L. de Haan and H. Ursprung, eds.), pp. 611–628. Holt, Rinehart and Winston, New York.

Jost, A. (1970a). General outline about reproductive physiology and its developmental background. *In* "Mammalian Reproduction" (H. Gibian and E. J. Platz, eds.), pp. 4–32. Springer Verlag, Berlin.

Jost, A. (1970b). Hormonal factors in the development of the male genital system. *In* "The Human Testis" (E. Rosenberg and C. A. Paulsen, eds.), pp. 11–17. Plenum, New York.

Jost, A. (1970c). Hormonal factors in sex differentiation. *Phil. Trans. Roy. Soc. London* **B259**, 119–130.

Jost, A., Chodkiewitz, M., and Mauleon, P. (1963). Intersexualité du foetus de veau prodnite par des androgènes. *C. R. Acad. Sci. (Paris)* **256**, 274–276.

Kaback, M. M., and Bernstein, L. H. (1970). Biologic studies of trisomic cells grown *in vitro*. *Ann. N.Y. Acad. Sci.* **171**, 526–536.

Kahn, J. (1962). The nucleolar organizer in the mitotic chromosome complement of *Xenopus laevis. Quart. J. Microscop. Sci.* **103**, 407–409.

Kallman, K. D. (1965). Genetics and geography of sex determination in the pocillid fish *Xiphophorus maculatus. Zoologica* **50**, 151–190.

Kallman, K. D. (1968). Evidence for the existence of transformer genes for sex in the teleost *Xiphophorus maculatus. Genetics* **60**, 811–828.

Kallman, K. D. (1970). Sex determination and the restriction of pigment pattern to the X and Y chromosomes in populations of a poeciliid fish, *Xiphophorus maculatus,* from the Sibun and Belize rivers of British Honduras. *Zoologica* **55**, 1–16.

Kaufmann, B. P. (1934). Somatic mitoses in *Drosophila melanogaster. J. Morphol.* **56**, 125–155.

Kawaguchi, K., and Marumo, R. (1967). Quoted by Yamamoto (1969).

Kesaree, N., and Woolley, P. V. (1963). A phenotypic female with 49 chromosomes, presumably XXXXX. *J. Pediat.* **63**, 1099–1103.

Keutel, J., Dollmann, A., and Munster, W. (1970). Triploidie (69,XXY) bei einem lebend geborenen Kind. *Z. Kinderheilk.* **109**, 104–117.

King, M. J., Gillis, E. M., and Baikie, A. G. (1962). Alkaline phosphatase activity of polymorphs in mongolism. *Lancet* ii, 1302–1305.

Kinsey, J. D. (1967). X-chromosome replication in early rabbit embryos. *Genetics* **55**, 337–343.

Kirk, D., and Jones, R. N. (1970). Nuclear genetic activity in B-chromosome of rye, in terms of the quantitative interrelationships between nuclear protein, nuclear RNA and histone. *Chromosoma* **31**, 241–254 (1970).

Klinefelter, H. F., Jr., Reifenstein, E. C., and Albright, F. (1942). Syndrome characterised by gynecomastia, aspermatogenesis without aleydigism, and increased excretion of follicle stimulating hormone. *J. Clin. Endocrinol. Metab.* **2**, 615–627.

Klinger, H. P., Davis, J., Goldhuber, P., and Ditta, T. (1968). Factor influencing mammalian X chromosome condensation and sex chromatin formation. I. The effect of *in vitro* cell density on sex chromatin frequency. *Cytogenetics* **7**, 39–57.

Kosswig, C. (1927). Über Bastarde der Teleostier *Platypoecilus* und *Xiphophorus. Z. Indukt. Abstamm. Vererbungsl.* **44**, 253.

Kosswig, C. (1931). Die geschlechtliche Differenzierung bei den *Bastarden* von *Xiphophorus helleri* und *Platypoecilus maculatus* und deren Nachkommen. *Z. Indukt. Abstamm. Vererbungsl.* **57**, 226–305.

Kosswig, C. (1932). Genotypische und phänotypische Geschlechtsbestimmung bei Zahnkarpfen und ihren Bastarden. I. *Z. Indukt. Abstamm. Vererbungsl.* **62**, 23–24.

Kosswig, C. (1964). Polygenic sex determination. *Experientia* **20**, 1–10.

Krooth, R. S. (1969). Gene action in human diploid cell strains. *In* "Comparative Mammalian Cytogenetics" (K. Benirschke, ed.), pp. 154–179. Springer Verlag, Berlin.

Landauer, W. (1960). Nicotine-induced malformations of chicken embryos and their bearing on the phenocopy problem. *J. Exp. Zool.* **143**, 107–122.

Landauer, W. (1965). Gene and phenocopy. Selection experiments and tests with 6-aminonicotinamide. *J. Exp. Zool.* **160**, 345–354.

Landauer, W., and Sopher, D. (1970). Succinate, glycerophosphate and ascorbate as sources of cellular energy and as antiteratogens. *J. Embryol. Exp. Morphol.* **24**, 187–202.

Lawler, S. D., and Sanger, R. (1970). Xg blood-groups and clonal—origin theory of chronic myeloid leukaemia. *Lancet* i, 584–585.

Lea, D. E. (1955). "Action of Radiations on Living Cells," 2nd ed. Cambridge Univ. Press, London.

Lederberg, J., and Tatum, E. L. (1946). Gene recombination in *E. coli. Nature (London)* **158**, 558.

Leisti *et al.* (1970). Cited by Chambon (1972).

Lewis, E. B. (1950). The phenomenon of position effect. *Advan. Genet.* **3**, 73–115.

Lennartz, K. J., Schümnelfeder, N., and Maurer, W. (1966). Dauer der DNS—Synthesephase bei Ascitestumoren der Maus unterschiedlicher Ploidie. *Naturwissenschaften* **53**, 21–22.

Lewis, E. B. (1964). Genetic control and regulation of developmental pathways. *In* "The Roles of Chromosomes in Development" (M. Locke, ed.), pp. 231–252. Academic Press, New York.

Lillie, F. R. (1917). The freemartin: A study of the action of sex hormones in the foetal life of cattle. *J. Exp. Zool.* **23**, 371–452.

Lillie, F. R. (1952). "Development of the Chick," 3rd ed. (Revised by H. L. Hamilton.) Henry Holt, New York.

Lima-de-Faria, A. (1959). Differential uptake of tritiated thymidine into hetero- and euchromatin in *Melanoplus* and *Secale. J. Biophys. Biochem. Cytol.* **6**, 457–466.

Lima-de-Faria, A. (ed.) (1969). DNA replication and gene amplification in heterochromatin. *In* "Handbook of Molecular Cytology," pp. 277–325. North Holland, Amsterdam.

Linder, D., and Gartler, S. M. (1965). Glucose-6-phosphate dehydrogenase mosaicism: utilization as a cell marker in the study of leiomyomas. *Science* **150**, 67–69.

Lindh, J. (1961). Quantitative Aspects of Prenatal Gonad Growth in Albino Rat and Golden Hamster. Studied by Morphologic and Experimental Methods. Department of Anatomy, Lund.

Lindsley, D. L., and Grell, E. H. (1968). Genetic Variation in *Drosophila melanogaster. Carnegie Inst. Wash. Publ. No.* 627.

Lindsten, J. (1963). "The Nature and Origin of X Chromosome Aberrations in Turner's Syndrome." Almqvist & Wiksell, Uppsala.

Liu, C. K. (1944). Quoted by Chan, S.T.H. (1970).

Lucas, M., and Dewhurst, C. J. (1972). Y chromosome fluorescence in phenotypic females. *J. Obstet. Gynaecol. Brit. Commonw.* **79**, 498–503.

Lyndon, R. F. (1968). The nucleus—its structure, function and development *In* "Plant Cell Organelles" (J. B. Pridham, ed.) pp. 16–39. Academic Press, New York.

Lyon, M. F. (1961). Gene action in the X-chromosome of the mouse. *Nature* (*London*) **190**, 372–373.

Lyon, M. F. (1970). Genetic activity of sex chromosomes in somatic cells of mammals. *Phil. Trans. Roy. Soc. London* **B259**, 41–52.

Lyon, M. F. (1971). Possible mechanisms of X-chromosome inactivation. *Nature New Biol.* **232**, 229–232.

Lyon, M. F. (1972). X-chromosome inactivation and developmental patterns in mammals. *Biol. Rev.* **47**, 1–35.

Lyon, M. F., and Hawkes, S. C. (1970). X-linked gene for testicular feminization in the mouse. *Nature* (*London*) **227**, 1217–1219.

McClintock, B. (1933). The association of non-homologous parts of chromosomes in the mid-prophase of meiosis in Zea mays. *Z. Zellforsch. Mikrosk. Anat.* **21**, 294–328.

McClintock, B. (1959). Genetic and cytological studies of maize. *Carnegie Inst. Wash. Yearb.* **58**, 452–456.

McClung, C. E. (1901). Notes on the accessory chromosome. *Anat. Anz.* **20**, 220–226.

McClung, C. E. (1902). The accessory chromosome—sex determinant? *Biol. Bull.* **3**, 43–84.

McGrady, E., Jr., (1938). The embryology of the opossum. *Amer. Anat. Memoirs No.* **16**, 1–233.

McFeely, R. A., Hare, W. C. D., and Biggers, J. D. (1967). Chromosome studies in 14 cases of intersex in domestic mammals. *Cytogenetics* **6**, 242–253.

McKusick, V. A. (1960). "Heritable Disorders of Connective Tissue," 2nd ed. Mosby Co., St. Louis.

McLaren, A., and Bowman, P. (1969). Mouse chimaeras derived from fusion of embryos differing by nine genetic factors. *Nature* (*London*) **224**, 238–240.

McLaren, A., Chandley, A. C., and Kofman-Alfaro, S. (1972). A study of meiotic germ cells in the gonads of foetal mouse chimaeras. *J. Embryol. Exp. Morphol.* **27**, 515–524.

Manolov, G., Manolova, Y., and Levan, A. (1971). The fluorescence pattern of the human karyotype. *Hereditas* **69**, 273–286.

Mather, K. (1944). The genetical activity of heterochromatin. *Proc. Roy. Soc.* **B132**, 308–332.

Mather, K., and Harrison, B. J. (1949). The manifold effect of selection. *Heredity* **3**, 1–52, 131–162.

Mather, K., and Jinks, J. L. (1971). "Biometrical Genetics," 2nd ed. Cornell Univ. Press, Ithaca, New York.

Matthey, R. (1950). Les chromosomes sexuels géants de *Microtus agrestis*. *Cellule* **53**, 163–184.

Matton-van Leuven, M. T., and François, J. (1970). Equine heterosexual chimeric twins: cytogenetic evaluation. *Bull. Europ. Soc. Human Genet.* **4**, 54–61.

Mellman, W. J., Younkin, L. H., and Baker, D. (1970). Abnormal lymphocyte function in trisomy 21. *Ann. N.Y. Acad. Sci.* **171**, 537–542.

Mendel, G. (1866). Versuche über Pflanzenhybride. English translation in "The Origin of Genetics" (C. Stern and E. R. Sherwood, eds.) pp. 1–48. Freeman, San Francisco, California (1966).

Mendel, G. (1870). See Stern and Sherwood, 1966.

Metz, C. W. (1914). Chromosome studies in the Diptera. I. A preliminary survey of the different types of chromosome groups in the genus. *J. Exp. Zool.* **17**, 45–59.

Mintz, B. (1964). Formation of genetically mosaic mouse embryos, and early development of "Lethal ($t^{12}t^{12}$)—Normal" mosaics. *J. Exp. Zool.* **157**, 273–292.

Mittwoch, U. (1967a). "Sex Chromosomes." Academic Press, New York.

Mittwoch, U. (1967b). Barr bodies in relation to DNA values and nuclear size in cultured human fibroblasts. *Cytogenetics* **6**, 38–50.

Mittwoch, U. (1967c). DNA synthesis in cells grown in tissue culture from patients with mongolism. *In* "Mongolism" (G. E. W. Wolstenholme and R. Porter, eds.), pp. 51–61. Churchill, London.

Mittwoch, U. (1967d). Sex differentiation in mammals. *Nature* (*London*) **214**, 554–556.

Mittwoch, U. (1969). Do genes determine sex? *Nature* (*London*) **221**, 446–448.

Mittwoch, U. (1970). How does the Y-chromosome affect gonadal differentiation? *Phil. Trans. Roy. Soc. London* **B259**, 113–117.

Mittwoch, U. (1971). Sex determination in birds and mammals. *Nature* (*London*) **231**, 432–434.

Mittwoch, U., and Buehr, M. L. (1973). Gonadal growth in embryos of *Sex reversed* mice. *Differentiation* (in press).

Mittwoch, U., and Delhanty, J. D. A. (1972). Inhibition of mitosis in human triploid cells. *Nature New Biol.* **238**, 11–13.

Mittwoch, U., Delhanty, J. D. A., and Beck, F. (1969). Growth of differentiating testes and ovaries. *Nature* (*London*) **224**, 1323–1325.

Mittwoch, U., Narayanan, T. L., Delhanty, J. D. A., and Smith, C. A. B. (1971). Gonadal growth in chick embryos. *Nature New Biol.* **231**, 197–200.

Monesi, V. (1965). Synthetic activities during spermatogenesis in the mouse. *Expl. Cell. Res.* **39**, 197–224.

Monesi, V. (1969). DNA, RNA, and protein synthesis during the mitotic cell cycle. *In* "Handbook of Molecular Cytology" (A. Lima-de-Faria, ed.), pp. 472–499. North Holland, Amsterdam.

Morgan, T. H. (1903). Recent theories in regard to the determination of sex. *Pop. Sci. Monthly* **64**, 97–116.

Morgan, T. H. (1910a). Chromosomes and heredity. *Amer. Natur.* **44**, 449–496.

Morgan, T. H. (1910b). Sex-limited inheritance in *Drosophila*. *Science* **32**, 120–122.

Morgan, T. H. (1926). "The Theory of the Gene." Yale Univ. Press, New Haven, Connecticut. Reprinted (1964) by Hafner, New York.

Morgan, T. H., and Bridges, C. B. (1916). Sex-linked inheritance in *Drosophila. Carnegie Inst. Wash. Publ.* 278, 1–122.

Morgan, T. H., and Goodale, H. D. (1912). Sex-linked inheritance in poultry. *Ann. N.Y. Acad. Sci.* 22, 113–133.

Morris, J. McL., (1953). The syndrome of testicular feminization in male pseudohermaphrodites. *Amer. J. Obstet. Gynecol.* 65, 1192–1211.

Morris, T. (1968). The XO and OY chromosome constitution in the mouse. *Genet. Res.* 12, 125–137.

Mukherjee, B. B., and Sinha, A. K. (1964). Single-active X-hypothesis; cytological evidence for random inactivation of X-chromosomes in a female mule complement. *Proc. Nat. Acad. Sci. U.S.* 51, 252–259.

Mukherjee, A. B., Moser, G. C., and Nitowsky, H. M. (1972). Fluorescence of X and Y chromatin in human interphase cells. *Cytogenetics* 11, 226–227.

Muller H. J. (1928). The production of mutations by X-rays. *Proc. Nat. Acad. Sci. U.S.* 14, 714–726.

Muller, H. J. (1930). Types of visible variations induced by X-rays in *Drosophia. J. Genet.* 22, 299–334.

Muller, H. J. (1932). Some genetic aspects of sex. *Amer. Natur.* 66, 118–138.

Müntzing, A. (1967). Some main results from investigations of accessory chromosomes. *Hereditas* 57, 432–438.

Mystkowska, E. T., and Tarkowski, A. K. (1968). Observations on CBA-p/CBA-T6 T6 mouse chimeras. *J. Embryol. Exp. Morphol.* 20, 33–52.

Naye, R. L. (1967). Prenatal organ and cellular growth with various chromosomal disorders. *Biol. Neonatorum* 11, 248–260.

Neel, J. V. (1949). The inheritance of sickle cell anaemia. *Science* 110, 64.

Neumann, F., Elger, W., and Steinbeck, H. (1970). Antiandrogens and reproductive development. *Phil. Trans. Roy. Soc. London* B259, 179–184.

Nilsson-Ehle, H. (1909). Kreuzungsuntersuchungen an Hafer and Weizen. *Acta Univ. Lund. Sect.* 2 5, 1–122.

Noble, C. K., and Kumpf, K. F. (1937). Sex reversal in the fighting fish. *Betta splendens. Anat. Rec.* 70 (Suppl. 1), 97.

Ohno, S. (1961). Sex chromosomes and microchromosomes of *Gallus domesticus. Chromosoma* 11, 484–498.

Ohno, S. (1970). "Evolution by Gene Duplication." Springer Verlag, Berlin.

Ohno, S. (1971). Simplicity of mammalian regulatory systems inferred by single gene determination of sex phenotypes. *Nature (London)* 234, 134–137.

Ohno, S., and Atkin, N. B. (1966). Comparative DNA values and chromosome complements of eight species of fishes. *Chromosoma* 18, 455–466.

Ohno, S., Kaplan, W. D., and Kinosita, R. (1959). Formation of the sex chromatin by a single X-chromosome in liver cells of *Rattus norvegicus. Exp. Cell. Res.* 18, 415–418.

Ohno, S., Trujillo, J. M., Kaplan, W. D., and Kinosita, R. (1961). Nucleolus organisers of chromosomal anomalies in man. *Lancet* ii, 123–126.

Ohno, S., Trujillo, J. M., Christian, L. C., and Teplitz, R. L. (1962). Possible germ cell chimeras among newborn dizygotic twin calves (*Bos taurus*). *Cytogenetics* 1, 258–265.

Ohno, S., Kitrell, W. A., Christian, L. C., Stenius, C., and Witt, G. A. (1963). An adult triploid chicken (*Gallus domesticus*) with a left ovotestis. *Cytogenetics* 2, 42–49.

Ohno, S., Tettenborn, U., and Dofuku, R. (1971). Molecular biology of sex differentiation *Hereditas* 69, 107–124.

Olby, R. C. (1966). "The Origins of Mendelism." Constable, London.

Overzier, C. (ed.) (1963a). The so-called true Klinefelter's syndrome. "Intersexuality," pp. 277–279. Academic Press, New York.

Overzier, C. (ed.) (1963b). True hermaphroditism. "Intersexuality," pp. 182–234. Academic Press, New York.

Overzier, C., and Hoffmann, K. (1963). Tumours with heterosexual activity. *In* "Intersexuality" (C. Overzier, ed.), pp. 402–461. Academic Press, New York.

Owen, J. J. T. (1965). Karyotype studies on *Gallus domesticus. Chromosoma* 16, 601–608.

Painter, T. S. (1924). The sex chromosomes of man. *Amer. Natur.* 58, 506–524.

Pardue, M. L., and Gall, J. C. (1970). Chromosomal localization of mouse satellite DNA. *Science* 168, 1356–1358.

Paris Conference (1973). *4th Int. Conf. Standardization in Human Cytogenetics, France, 1971* (J. L. Hamerton, P. A. Jacobs, and H. P. Klinger, eds.). National Foundation, March of Dimes.

Park, W. W. (1957). The occurrence of sex chromatin in early human and macaque embryos. *J. Anat.* 91, 369–373.

Paulmier F. C. (1899). The spermatogenesis of *Anasa tristis. J. Morphol. Suppl.* 15, 224–268.

Pearson, K. (1904a). Mathematical contribution to the theory of evolution XII. On a generalised theory of alternative inheritance, with special reference to Mendel's law. *Phil. Trans. Roy. Soc. London* A203, 53–86.

Pearson, K. E. (1904b). On the laws of inheritance in man. II. On the inheritance of the mental and moral characters in man, and its comparison with the inheritance of physical characters. *Biometrika* 3, 131–190.

Pearson, P. L., Bobrow, M., and Vosa, C. G. (1970). Technique for identifying Y chromosomes in human interphase nuclei. *Nature* (*London*) 226, 78–80.

Pearson, P. L., Bobrow, M., Vosa, C. G., and Barlow, P. W. (1971). Quinacrine fluorescence in mammalian chromosomes. *Nature* (*London*) 231, 326–329.

Penrose, L. S. (1951). Measurement of pleiotropic effects in phenylketonurea. *Ann. Eugen.* 16, 134–141.

Penrose, L. S. (1957). Genetics of anencephaly. *J. Ment. Def. Res.* 1, 4-15.

Penrose, L. S. (1967). Finger-print pattern and the sex chromosomes. *Lancet* i, 298–300.

Penrose, L. S. (1969). Dermatoglyphics. *Sci. Amer.* 221, 72–83.

Penrose, L. S., and Smith, G. F. (1966). "Down's Anomaly." Churchill, London.

Peters, H. (1970). Migration of gonocytes into the mammalian gonad and their differentiation. *Phil. Trans. Roy. Soc. London* **B259**, 91–101.

Pierce, W. P. (1937). The effect of phosphorus on chromosome and nuclear volume in a violet species. *Bull. Torrey Bot. Cl.* **64**, 345–354.

Piko, L., and Bomsel-Helmreich, O. (1960). Triploid rat embryos and other chromosomal deviations after colchicine treatment and polyspermy. *Nature (London)* **186**, 737–739.

Plate, L. (1910). See Plate, L. (1932). "Vererbungslehre," Vol. I. Fischer, Jena.

Plaut, W. S., and Mazia, D. (1956). The distribution of newly synthesized DNA in mitotic division. *J. Biophys. Biochem. Cytol.* **2**, 573–588.

Polani, P. E. (1969). Abnormal sex chromosomes and mental disorder. *Nature (London)* **223**, 680–686.

Polani, P. E. (1970). Hormonal and clinical aspects of hermaphroditism and the testicular feminizing syndrome in man. *Phil. Trans. Roy. Soc. London* **B259**, 187–204.

Polani, P. E., Angell, R., Gianelli, F., de la Chapelle, A., Race, R. R., and Sanger, R. (1970). Evidence that the $Xg$ locus is inactivated in structurally abnormal X chromosomes. *Nature (London)* **227**, 613–616.

Pratt, J., Sarret, E., Moragas, A., and Martin, C. (1971). Triploid live full term infant. *Helv. Paediat. Acta* **2**, 164–171.

Price, D. (1970). *In vitro* studies on differentiation of the reproductive tract. *Phil. Trans. Roy. Soc. London* **B259**, 133–139.

Price, D., and Ortiz, E. (1965). The role of fetal androgen in sex differentiation in mammals. *In* "Organogenesis" (R. L. de Haan and H. Ursprung, eds.), pp. 629–562. Holt, Rinehart & Winston, New York.

Punnett, R. C., and Bailey, P. G. (1914). On inheritance of weight in poultry. *J. Genet.* **4**, 23–39.

Quetelet, A. (1871). "Anthropometrie." Muquard, Brussels.

Race, R. R. (1973). Is the Xg blood group subject to inactivation? *Proc. 4th Congr. Human Genet., Paris, 1971.* Excerpta Medica, Amsterdam. (in press).

Race, R. R., and Sanger, R. (1968). "Blood Groups in Man" 5th ed. Blackwell, Oxford.

Rasch, E. M., Barr, H. J., and Rasch, R. W. (1971). The DNA content of sperm of *Drosophila melanogaster*. *Chromosoma* **33**, 1–18.

Ratazzi, M. C., and Cohen, M. M. (1972). Further proof of genetic inactivation of the X chromosome in the female mule. *Nature (London)* **237**, 393–395.

Ratcliffe, S., Stewart, A. L., Melville, M. M., Jacobs, P. A., and Keay, A. J. (1970). Chromosome studies on 3500 newborn male infants. *Lancet* **i**, 121–122.

Ray, M., Gee, P. A., Richardson, B. J., and Hamerton, J. L. (1972). G6PD expression and X chromosome late replication in fibroblast clones from a female mule. *Nature (London)* **237**, 396–397.

Ray-Chaudhuri, S. P., Singh, L., and Sharma, T. (1970). Sexual dimorphism in somatic interphase nuclei of snakes. *Cytogenetics* **9**, 410–423.

Renwick, J. H. (1972). Hypothesis: anencephaly and spina bifida are usually preventable by avoidance of a specific but unidentified substance present in certain potato tubers. *Brit. J. Prev. Soc. Med.* **26**, 67–88.

Richardson, B. J., Czuppon, A. B., and Sharman, G. B. (1970). Inheritance of glucose-6-phosphate dehydrogenase variation in kangaroos. *Nature New Biol.* **230**, 154–155.

Ritossa, F. M., and Spiegelman, S. (1965). Localization of DNA complementary to ribosomal RNA in the nucleolus organizer region of *Drosophila melanogaster*. *Proc. Nat. Acad. Sci. U.S.* **53**, 737–745.

Ritossa, F. M., Atwood, K. C., and Spiegelman, S. (1966). A molecular explanation of the bobbed mutants of Drosophila as partial deficiencies of "ribosomal" DNA. *Genetics* **54**, 819–834.

Röhrborn, G. (1970). Biochemical mechanisms of mutations. *In* "Chemical Mutagenesis in Mammals and Man" (F. Vogel & G. Röhrborn, eds.), pp. 1–15. Springer Verlag, Berlin.

Romanoff, A. L. (1960). "The Avian Embryo: Structural and Functional Development." Macmillan, New York.

Rook, A., Hsu, L. Y., Gertner, M., and Hirschhorn, K. (1971). Identification of Y and X chromosomes in amniotic fluid cells. *Nature* (*London*) **230**, 53.

Rowley, J. D., Muldal, S., Gilbert, C. W., Lajtha, L. G., Lindsten, J., Fraccaro, M., and Kajser, K. (1963). Synthesis of deoxyribonucleic acid on X-chromosomes of an XXXXY male. *Nature* (*London*) **197**, 251–252.

Rowley, J. D., and Bodmer, W. F. (1971). Relationship of centromeric heterochromatin to fluorescent banding patterns of metaphase chromosomes in the mouse. *Nature* (*London*) **231**, 503–506.

Russell, L. B., and Chu, E. H. Y. (1961). An XXY male in the mouse. *Proc. Nat. Acad. Sci. U.S.* **47**, 571–575.

Russell, W. L., Russell, L. B., and Gower, J. S. (1959). Exceptional inheritance of a sex-linked gene in the mouse explained on the basis that the X/O sex chromosome constitution is female. *Proc. Nat. Acad. Sci. U.S.* **45**, 554–560.

Rutter, W. J., Clark, W. R., Kemp, J. D., Bradshaw, W. S., Sanders, T. S., and Ball, W. D. (1968). Multiphasic regulation in cytodifferentiation. *In* "Epithelial-Mesenchymal Interactions" (R. Fleischmajer and R. E. Billingham, eds.), pp. 114–131. Williams & Wilkins, Baltimore, Maryland.

Sandquist, U., and Hellström, E. (1969). Transmission of 47, XYY karyotypes. *Lancet* **ii**, 1367.

Sanger, R., Tippett, P., and Gavin, J. (1971). Xg groups and sex abnormalities in people of Northern European ancestry. *J. Med. Genet.* **8**, 417–426.

Saunders, G. F., Hsu, T. C., Getz, M. J., Simes, E. L., and Arrighi, F. E. (1972). Locations of human satellite DNA in human chromosomes. *Nature, New Biol.* **236**, 244–246.

Saxén, L., and Rapela, J. (1969). "Congenital Defects." Holt, Rinehart and Winston, New York.

Schindler, A-M., and Mikamo, K. (1970). Triploidy in man. *Cytogenetics* **9**, 116–130.

Schmickel, R. D., Silverman, E. M., Floy, A. D., Payne, F. E. Pooley, J. M. and Beck, M. L. (1971). A liveborn infant with 69 chromosomes. *J. Pediat.* **79**, 97–103.

Schmid, W. (1962). DNA replication patterns of the heterochromosomes in *Gallus domesticus. Cytogenetics* **1**, 344–352.

Schmid, W. (1963). DNA replication patterns of human chromosomes. *Cytogenetics (Basel)* **2**, 175–193.

Schmid, W. (1967). Heterochromatin in mammals. *Arch. Julius Klaus-Stift. Vererbungsforsch. Sozialanthropol. Rassenhyg.* **42**, 1–60.

Schmid, W., and Leppert, M. F. (1968). Karyotyp, Heterochromatin und DNS—Werte bei 13 Arten von Wühlmäusen (Microtinae, Mammalia-Rodentia). *Arch. Julius Klaus-Stift. Vererbungsforsch. Sozialanthropol. Rassenhyg.* **43**, 88–91.

Schmid, W., Smith, D. W., and Theiler, K. (1965). Chromatinmuster in verschiedenen Zelltypen und Lokalisation von Heterochromatin auf Metaphasenchromosomen bei *Microtus agrestis, Mesocricetus auratus, Cavia cobaya* und beim Menschen. *Arch. Julius Klaus-Stift. Vererbungsforsch. Sozialanthropol. Rassenhyg.* **40**, 35–49.

Schnedl, W. (1971). Analysis of the human karotype using a reassociation technique. *Chromosoma* **34**, 448–454.

Schnedl, W., and Schnedl, M. (1972). Banding patterns in rat chromosomes (*Rattus norvegicus*). *Cytogenetics* **11**, 188–196.

Schrader, F. (1928). "Die Geschlechtschromosomen." Borntraeger, Berlin.

Schrader, F. (1929). Experimental and cytological investigations of the life-cycle of Gossyparia spuria (Coccidae) and their bearing on the problem of haploidy in males. *Z. Wiss. Zool.* **138**, 386–408.

Schwarzacher, H. G., and Pera, F. (1970). Das Problem der Sexchromatinnegativen Zellen. *Z. Anat. Entwicklungsgesch.* **132**, 18–33.

Seabright, M. (1972). The use of proteolytic enzymes for the mapping of structural rearrangements in the chromosomes of man. *Chromosoma* **36**, 204–210.

Seiler, J. (1914). Das Verhalten der Geschlechtschromosomen bei Lepidopteren. *Arch. Zellforsch.* **13**, 159–259.

Sharman, G. B. (1971). Late DNA replication in the paternally derived X chromosome of female kangaroos. *Nature (London)* **230**, 231–232.

Short, R. V. (1970). The bovine freemartin: A new look an old problem. *Phil. Trans. Roy. Soc. London* **B259**, 141–147.

Short, R. V., Smith, J., Mann, T., Evans, E. P., Hallett, J., Fryer, A., and Hamerton, J. L. (1969). Cytogenetic and endocrine studies of a freemartin heifer and its bull co-twin. *Cytogenetics* **8**, 369–388.

Siciliano, M. J., Perlmutter, A., and Clark, E. (1971). Effect of sex on the development of melanoma in hybrid fish of the genus *Xiphophorus. Cancer Res.* **31**, 725–729.

Simpson, J. L., Discher, R., Morillo-Cucci, G., and Conolly, C. E. (1972). Triploidy (69,XXY) in a liveborn infant. *Ann. Genet.* **15**, 103–106.

Singh, R. P., and Carr, D. H. (1966). The anatomy and histology of XO human embryos and fetuses. *Anat. Rec.* **155**, 369–383.

Smith, S. G. (1944). The diagnosis of sex by means of heteropycnosis. *McGill Med. J.* **13**, 451–456.

Sohval, A. R. (1964). Hermaphroditism with "atypical" or "mixed" gonadal dysgenesis. *Amer. J. Med.* **36**, 281–292.

Southern, D. I. (1970). Polymorphism involving heterochromatic segments in *Metrioptera brachyptera. Chromosoma* **30**, 154–168.

Southern, E. M. (1970). Base sequence and evolution of guinea-pig α- satellite D.N.A. *Nature (London)* **227**, 794–798.

Spillman, W. J. (1908). Spurious allelomorphism: Results of some recent investigations. *Amer. Natur.* **42**, 610–615.

Stern, C. (1929). Untersuchungen über Aberrationen des Y-chromosoms von *Drosophila melanogaster. Z. Indukt. Abstamm. Vererbungsl.* **51**, 252–353.

Stern, C. (1960). "Principles of Human Genetics." Freeman, San Francisco, California.

Stern, C. (1967). Genes and people. *Perspect. Biol. Med.* **10**, 500–523.

Stern, C., and Sherwood, E. R. (1966). "The Origin of Genetics." A Mendel Source Book. Freeman, San Francisco, California.

Stevens, N. M. (1905). Studies in spermatogenesis, with especial reference to the accessory chromosome. *Carnegie Inst. Wash. Publ.* **36**, 1–32.

Stevens, N. M. (1906). Studies in spermatogenesis. II. A comparative study of the heterochromosomes in certain species of Coleoptera, Hemiptera and Lepidoptera, with especial reference to sex determination. *Carnegie Inst. Wash. Publ.* **36**, 33–74.

Strasburger, E. (1900). Versuche mit diöcischen Pflanzen in Rücksicht auf Geschlechtsverteilung. *Biol. Zentrabl.* **20**, 657–665, 689–698, 721–731, 753–785.

Sturtevant, A. H. (1913). The linear arrangement of six sex-linked factors in *Drosophila,* as shown by their mode of association. *J. Exp. Zool.* **12**, 499–518.

Sturtevant, A. H. (1945). A gene in *Drosophila melanogaster,* that transforms females into males. *Genetics* **30**, 297–299.

Sturtevant, A. H. (1965). "A History of Genetics." Harper & Row, New York.

Sumner, A. T., Robinson, J. A., and Evans, H. J. (1971a). Distinguishing between X, Y and YY—bearing human spermatozoa by fluorescence and DNA content. *Nature New Biol.* **229**, 231–233.

Sumner, A. T., Evans, H. J., and Buckland, R. A. (1971b). New technique for distinguishing between human chromosomes. *Nature New Biol.* **232**, 31–32.

Sutton, W. S. (1902). On the morphology of the chromosome group of *Brachystola magna. Biol. Bull.* **4**, 24–39.

Sutton, W. S. (1903). The chormosomes in heredity. *Biol. Bull.* **4**, 231–251.

Swift, C. H. (1915). Origin of the definitive sex-cells in the female chick and their relation to the primordial germ-cells. *Amer. J. Anat.* **18**, 441–470.

Swift, H. (1950). The constancy of desoxyribonucleic acid in plant nuclei. *Proc. Nat. Acad. Sci. U.S.* **36**, 643–654.

Swyer, G. I. M. (1955). Male pseudohermaphroditism: a hitherto undescribed form. *Brit. Med. J.* **ii**, 709–712.

Tandler, J., and Keller, K. (1911). Über das Verhalten des Chorions bei verschieden-geschlechtlicher Zwillingsgravidität des Rindes und über die Morphologie des Genitales der weiblichen Tiere, welche einer solchen Gravidität entstammen. *Deut. Tieraerztl. Wochenschr.* **19**, 148–153.

Tanner, J. M. (1962). "Growth at Adolescence," 2nd ed. Blackwell, *Publ.* Oxford.

Tanner, J. M. (1962). "Growth at Adolescence," 2nd ed. Blackwell, *Publ.* Oxford.

Tarkowski, A. K. (1961). Mouse chimeras developed from fused eggs. *Nature* (*London*) **190**, 857–860.

Tarkowski, A. K. (1970). Germ cells in natural and experimental chimaeras in mammals. *Phil. Trans. Roy. Soc. London* **B259**, 107–111.

Taylor, A. I. (1968). Autosomal trisomy syndromes: a detailed study of 27 cases of Edward's syndrome and 27 cases of Patau's syndrome. *J. Med. Genet.* **5**, 227–252.

Taylor, J. H. (1960). Asynchronous duplication of chromosomes in cultured cells of Chinese hamster. *J. Biophys. Biochem. Cytol.* **7**, 455–464.

Taylor, J. H. (1965). "Selected Papers on Molecular Genetics." Academic Press, New York.

Taylor, J. H., Woods, P. S., and Hughes, W. L. (1957). The organization and duplication of chromosomes as revealed by autoradiographic studies. *Proc. Nat. Acad. Sci. U.S.* **43**, 122–128.

Teter, J., and Boczkowski, K. (1967). Occurrence of tumors in dysgenetic gonads. *Cancer* **20**, 1301–1310.

Therkelsen, A. J., and Petersen, G. B. (1962). Frequency of sex-chromatin positive cells in the logarithmic and post-logarithmic growth phase of human cells in tissue culture. *Exp. Cell. Res.* **28**, 588–590.

Thorburn, M. J. (1964). Sex-chromatin in a 13-day embryo. *Lancet* **i**, 277–278.

Thuline, H. C., and Norby, D. E. (1961). Spontaneous occurrence of chromosome abnormality in cats. *Science* **134**, 554.

Tiepolo, L., Fraccaro, M., Hultén, M., Lindsten, J., Mannini A., and Ming, P. M. L. (1967). Timing of sex chromosome replication in somatic and germ-like cells of the mouse and the rat. *Cytogenetics* **6**, 51–66.

Tjio, J. H., and Levan, A. (1956). The chromosome number of man. *Hereditas* **42**, 1–6.

Tjio, J. H., Puck, T. T., and Robinson, A. (1960). The human chromosomal satellites in normal persons and in two patients with Marfan's syndrome. *Proc. Nat. Acad. Sci. U.S.* **46**, 532–539.

Tomkins, G. M., Gelehrter, T. D., Granner, D., Martin, D., Samuels, H. H., and Thompson, E. B. (1969). Control of specific gene expression in higher organisms. *Science* **166**, 1474.

Troy, M. R., and Wimber, D. E. (1968). Evidence for a constancy of the

DNA synthetic period between diploid/polyploid groups in plants. *Exp. Cell Res.* **53**, 145–154.

Tsaney, R., and Sendov, Bl. (1971). Possible molecular mechanism for cell differentiation in multicellular organisms. *J. Theoret. Biol.* **30**, 337–393.

Turner, H. H. (1938). A syndrome of infantilism, congenital webbed neck and cubitus valgus. *Endocrinology* **23**, 566–574.

Utakoji, T., and Hsu, T. C. (1965). DNA replication patterns in somatic and germline cells of the male Chinese hamster. *Cytogenetics* **4**, 295–315.

van't Hof, J. (1965). Relationship between mitotic cycle duration, S period duration and the average rate of DNA synthesis in the root meristem of cells of several plants. *Exp. Cell Res.* **39**, 48–58.

van Tienhoven, A. (1961). Endocrinology of reproduction in birds. *In* "Sex and Internal Secretions" (W. C. Young ed.), 3rd ed., Vol. 2, pp. 1088–1169. Ballière, Tindall & Cox, London.

van Wagenen, G. and Simpson, M. E. (1965). "Embryology of the Ovary and Testis *Homo sapiens* and *Macaca mulatta*." Yale Univ. Press, New Haven, Connecticut.

Villee, C. A. (1942). A study of hereditary homoeosis: the mutant tetraltera in *Drosophila melanogaster*. *Univ. Calif. Pub. Zool.* **49**, 125–183.

Villee, C. A. (1945). Developmental interactions of homeotic and growth rate genes in *Drosophila melanogaster*. *J. Morphol.* **77**, 105–118.

Villee, C. A. (1961). Some problems of the metabolism and mechanism of action of steroid sex hormones. *In* "Sex and Internal Secretions" (W. C. Young, ed.), 3rd ed., pp. 643–665. Baillière, Tindall & Cox, London.

Vogel, F. (1964). A preliminary estimate of the number of human genes. *Nature (London)* **201**, 847.

Vogel, F. (1970). Spontaneous mutation in man. *In* "Chemical Mutagenesis in Mammals and Man" (F. Vogel and G. Röhrborn, eds.), pp. 16–68. Springer Verlag, Berlin.

Vosa, C. G. (1970). The discriminating fluorescence patterns of the chromosomes of *Drosophila melanogaster*. *Chromosoma* **31**, 446–451.

von Wettstein, F. (1924). Über Fragen der Geschlechtsbestimmung bei Pflanzen. *Naturwissenschaften* **12**, 761.

Waddington, C. H. (1953). Genetic assimilation of an acquired character. *Evolution* **7**, 118–126.

Waddington, C. H. (1961). Genetic assimilation. *Advan. Genet.* **10**, 257–293.

Wahrman, J. (1972). Private communication.

Walker, P. M. B., Flamm, W. G., and McLaren, A. (1969). Highly repetitive DNA in rodents. *In* "Handbook of "Molecular Cytology" (A. Lima-de-Faria, ed.), pp. 52–66. North Holland, Amsterdam.

Warmke, H. E. (1946). Sex determination and sex balance in *Melandrium*. *Amer. J. Bot.* **33**, 648–660.

Watson, J. D. (1970). "Molecular Biology of the Gene," 2nd ed. Benjamin, New York.

Watson, J. D., and Crick, F. H. C. (1953). A structure for deoxyribonucleic acid. *Nature (London)* **171**, 737–738.

Weatherall, D. J., and Clegg, J. B. (1969). Disorders of protein synthesis. In "Selected Topics in Medical Genetics" (C. A. Clarke, ed.), pp. 110–134. Oxford University Press, London.

Weber, W. W., Mittwoch, U., and Delhanty, J. D. A. (1965). Leucocyte alkaline phosphatase in Klinefelter's syndrome. *J. Med. Genet.* **2**, 112–115.

Wells, L. J. (1962). Experimental studies on the role of the developing gonads in mammalian sex differentiation. *In* "The Ovary" (S. Zuckerman, ed.), Vol. 2, pp. 131–153. Academic Press, New York.

Welshons, W. J., and Russell, L. B. (1959). The Y-chromosome as the bearer of the male determining factors in the mouse. *Proc. Nat. Acad. Sci. U.S.* **45**, 560–566.

Wessels, N. K. (1968). Problems in the analysis of determination, mitosis and differentiation. *In* "Epithelial-Mesenchymal Interactions" (R. Fleischmajer and R. E. Billingham, eds.), pp. 132–151. Williams & Wilkins, Baltimore, Maryland.

Westergaard, M. (1958). The mechanism of sex determination in dioecious flowering plants. *Advan. Genet.* **9**, 217–281.

White, M. J. D. (1954). "Animal Cytology and Evolution." 2nd ed. Cambridge Univ. Press, London.

Whitehouse, H. L. K. (1969). "Towards an Understanding of the Mechanism of Heredity," 2nd ed. Arnold, London.

Wildermuth, H. (1970). Determination and transdetermination in cells of the fruitfly. *Sci. Progr.* **58**, 329–358.

Wilkins, L. (1965). "The Diagnosis and Treatment of Endocrine Disorders in Childhood and Adolescence," 3rd ed. Thomas, Springfield, Illinois.

Wilkins, L., and Fleischman, W. (1944). Ovarian agenesis; pathology, associated clinical symptoms and the bearing on the theories of sex differentiation. *J. Clin. Endocrinol.* **14**, 1270–1271.

Wilkins, M. H. F., Stokes, A. R., and Wilson, H. R. (1953). Molecular structure of deoxypentonucleic acid. *Nature (London)* **171**, 738–740.

Wilson, E. B. (1896). "The Cell in Development and Inheritance," 1st ed. Macmillan, New York.

Wilson, E. B. (1905). Studies on chromosomes. I. The behaviour of the idiochromosomes in Hemiptera. *J. Exp. Zool.* **2**, 371–405.

Wilson, E. B. (1909). Studies on chromosomes. IV. The "accessory" chromosome in *Syromastes* and *Pyrrhocoris* with a comparative review of the type of sexual differences of the chromosome group. *J. Exp. Zool.* **6**, 69–99.

Wilson, E. B. (1911). The sex chromosomes. *Arch. Mikrosk. Anat. Entwicklungsmech.* **77**, II, 249–271.

Winge, O. (1922a). A peculiar mode of inheritance and its cytological explanation. *J. Genet.* **12**, 137–144.

Winge, O. (1922b). One-sided masculine and sex-linked inheritance in *Lebistes reticulatus*. *J. Genet.* **12**, 145–162.

Winge, O. (1927). The location of eighteen genes in *Lebistes reticulatus*. *J. Genet.* **18**, 1–43.

Winge, O. (1932). The nature of sex chromosomes. *Proc. 6th Int. Congr. Genet. Ithaca, New York, 1932,* Vol. 1, pp. 343–355. Brooklyn Botanic Gardens, Brooklyn, New York.

Winge, O. (1934). The experimental alteration of sex chromosomes into autosomes, and vice versa, as illustrated by *Lebistes. C.R. Lab. Carlsberg Ser. Physiol.* **21**, 1–49.

Winge, O. (1937). Goldschmidt's theory of sex determination in *Lymantria. J. Genet.* **34**, 31–39.

Winge, O., and Ditlevsen, E. (1938). A lethal gene in the Y chromosome of *Lebistes. C.R. Lab. Carlsberg, Ser. Physiol.* **22**, 205–211.

Winge, O., and Ditlevson, E. (1947). Colour inheritance and sex determination in *Lebistes. Heredity* **1**, 65–83.

Winge, O. and Ditlevsen, E. (1948). Colour inheritance and sex determination *Lebistes. C. R. Trav. Lab. Carlsberg, Ser. Physiol.* **24**, 227–248.

Witschi, E. (1935). The chromosomes of hermaphrodites. *Biol. Bull.* **68**, 263–267.

Witschi, E. (1942). Temperature factors in the development and evolution of sex. *Biol. Symp.* **6**, 51–70.

Witschi, E. (1948). Migration of the germ cells of human embryos from the yolk sac to the primitive gonadal folds. *Contrib. Embryol. Carnegie Inst. Wash.* **32**, 67–80.

Witschi, E. (1956). "Development of Vertebrates." Saunders, Philadelphia, Pennsylvania.

Wolf, U., Flinspach, G., Bohm, R., and Ohno, S. (1965). DNA—Reduplikationsmuster bei den Riesengeschlechtschromosomen von *Microtus agrestis. Chromosoma* **16**, 609–619.

Wolff, E. (1962). Experimental modification of ovarian development. *In* "The Ovary" (S. Zuckerman, ed.), Vol. 2, pp. 81–129. Academic Press, New York.

Wolff, Et., and Wolff, Em. (1951). The effects of castration on bird embryos. *J. Exp. Zool.* **116**, 59–97.

Wright, S. (1934a). An analysis of variability in number of digits in an inbred strain of guinea pigs. *Genetics* **19**, 506–536.

Wright, S. (1934b). The results of crosses between inbred strains of guinea pigs differing in number of digits. *Genetics* **19**, 537–551.

Wright, S. (1968). "Evolution and the Genetics of Populations." Univ. Chicago Press, Chicago, Illinois.

Yamamoto, T. (1953). Artificially induced sex-reversal in genotypic males of the medaka (*Oryzias latipes*). *J. Exp. Zool.* **123**, 571–594.

Yamamoto, T. (1955). Progeny of artificially induced sex-reversals of male genotype (XY) in the medaka (*Oryzias latipes*) with special reference to YY-male. *Genetics* **40**, 406–419.

Yamamoto, T. (1958). Artificial induction of functional sex-reversal in genotypic females of the medaka (*Oryzias latipes*). *J. Exp. Zool.* **137**, 227–263.

Yamamoto, T. (1964). The problem of viability of YY zygotes in the medaka, *Oryzias latipes. Genetics* **50**, 45–58.

Yamamoto, T. (1967). Estrone-induced white YY females and mass production of white YY males in the medaka, *Oryzias latipes. Genetics* **55**, 329–336.

Yamamoto, T. (1969). Sex differentiation. *In* "Fish Physiology" (W. S. Hoar and D. J. Randall, eds.), Vol. 3, pp. 117–175. Academic Press, New York.

Yamamoto, T., and Matsuda, N. (1963). Effects of estradiol, stilbestrol and some alkyl-carbonyl androstanes upon sex differentiation in the medaka, *Oryzias latipes. Gen. Comp. Endocrinol.* **3**, 101–110.

Yamamoto, T., Takeuchi, K., and Takai, M. (1968). Male inducing action of androsterone and testosterone proprinonate upon XX zygotes in the medaka, *Oryzias latipes. Embryologia* **10**, 142–151.

Yang, D-P., and Dodson, E. O. (1970). The amounts of nuclear DNA and the duration of DNA synthetic period (S) in related diploid and auto-tetraploid species of oats. *Chromosoma* **31**, 309–320.

Yerushalmi, J. (1972). Infants with low birth weights born before their mothers started to smoke cigarettes. *Amer. J. Obstet. Gynec.* **112**, 277–284.

Young, W. C. (1961). The mammalian ovary. *In* "Sex and Internal Secretions" (W. C. Young, ed.), 3rd ed. Vol. 1, pp. 449–496. Baillière, Tindall and Cox, London.

Zaleski, W. A., Houston, C. S., Poszonyi, J., and Ying, K. L. (1966). The XXXXY chromosome anomaly: report of three cases and review of 30 cases from the literature. *Can. Med. Ass. J.* **94**, 1143–1154.

Zander, J., and Henning, H. D. (1963). Hormones and intersexuality. *In* "Intersexuality" (C. Overzier, ed.), pp. 118–172. Academic Press, New York.

Zech, L. (1969). Investigation of metaphase chromosomes with DNA-binding fluorochromes. *Exp. Cell Res.* **58**, 463.

# Author Index

Numbers in italics refer to the pages on which the complete references are listed.

Deys, B. B., 64, *208*
Diasco, R. B., 145, *208*
Dickson, J. M., 161, *213*
Diczfalusy, E., 131, *208*
Discher, R., 155, *225*
Ditlevsen, E., 93, 95, *230*
Ditta, T., 63, *217*
Dobzhansky, T., 26, 51, 52, 56, 165, *204, 208*
Dodson, E. O., 181, *231*
Dofuku, R., 193, 194, *222*
Dolfini, S., 19, 71, *202,* 146, *210*
Doll, R., 147, *216*
Dollmann, A., 155, *217*
Doncaster, L., 13, 15, *208*
Doniach, L., 70, *208*
Drets, M. E., 79, *208*
Düsing, C., 2, *208*
Dunn, H. G., 141, *203*
Dunn, L. C., 43, 45, *208*
Dzwillo, M., 95, *208*

**E**

East, E. M., 33, *208*
East, L. M., 33, *208*
Edwards, J. H., 39, 63, 155, 185, 186, *208*
Elger, W., 116, *221*
Ellison, J. R., 77, 186, *202, 208*
Elsdale, T. R., 56, *208*
Emerson, R. A., 33, *208*
Epifanova, O. I., 199, *208*
Evans, E. P., 144, *209,* 161, *225*
Evans, H. J., 76, 77, 79, 186, *209, 210, 226*

**F**

Fahmy, M. J., 54, *209*
Fahmy, O. G., 54, *209*
Falconer, D. S., 34, *209*
Fankhauser, C., 181, *209*
Farnsworth, M. W., 54, *209*
Fedrick, J., 169, *209*

Ferrier, P., 156, *209*
Ferrier, S., 156, *209*
Ferguson-Smith, M. A., 147, *209*
Fincham, J. R. S., 60, *209*
Fischberg, M., 56, *208,* 181, *209*
Fishelson, L., 92, *209*
Fitzgerald, P. H., 149, *209*
Flamm, W. G., 57, *209, 228*
Fleischman, W., 143, *229*
Flinspach, G., 64, *230*
Flory, A. D., 155, *224*
Förster, W., 96, *201*
Foley, G. E., 74, 77, *206*
Foote, C. L., 103, *209*
Forbes, T. R., 91, *210*
Ford, C. E., 137, 143, 144, 145, 158, *209, 210*
Forrest, H., 153, *216*
Forteza, G., 150, *210*
Fraccaro, M., 19, 71, 73, 141, 146, *202, 210, 224, 227*
François, J., 114, *219*
Fraser, F. C., 39, *210*
Fréderic, J., 15, *210*
Fredga, K., 157, *210*
Fryer, A., 68, *213,* 161, *225*

**G**

Galbis, M., 150, *210*
Gall, J. C., 79, *222*
Galton, F., 30–32, *210*
Ganner, E., 77, *210*
Garrod, A. E., 45, *210*
Gartler, S. M., 66, 148, *210, 218*
Gavin, J., 147, *224*
Gee, P. A., 68, *213, 223*
Geisler, M., 143, *210*
Gelehrter, T. D., 190, *227*
George, D. P., 146, *210*
German, J. L., 71, 73, 151, 156, *211*
Gertner, M., 76, *224*
Getz, M. J., 79, *224*

# Subject Index

## A

A chromosomes, 58
*ABO* locus, 66
Abortions, spontaneous, 143, 155
*Abraxas grossulariata,* 13–15
Accessory chromosome, 5, 6, 9
Accessory chromosomes, *see* B
  chromosomes
3-Acetylpyridine, 169
Acid phosphatase, 36
Acrocentric, 18, 190
Adenine, 46, 47, 77
Actinomycin, 179
Adrenal cortex, 198
  tumors in, 154
Adrenogenital syndrome, 154
*Aedes stimulans,* 133
Age
  of egg, sperm, 2
  and loss of Y chromosome, 148
  of mother, father, 2
Albinism, 45
Alcohol dehydrogenase, 153, 194
Alkaptonuria, 45
Allele, 31, 33, 36, 45, 56, 64, 154,
  171, 194
Allelomorph, 17, 31
Allosome, 9
Amenorrhea
  and ovarian tumors, 154
  in XXX females, 143
Amino acids, 47, 74
6-Aminonicotinamide, 169, 171
Amphibia, 91, 103, 105, 155, 181,
  196
  DNA in, 189
*Amphioxus lanceolatus,* 189
*Anasa tristis,* 5
Androgens, 88, 104, 154, 160, 161,
  196, 198
Androstenedione, 198
Anencephaly, 39, 170
Aneuploidy, 119
Anlage, 45

Antiandrogen, 116
Antimetabolites, 169
*Antirrhinum,* 60
*Anthias squamipinnis,* 92
Aphids, 16
*Aplocheilus, see Oryzias*
*Aristapedia,* 165, 166, 168
Ascites tumor, 182
Ascorbate, 169
Autoradiography, 50, 68, 70–74
Autosomal locus
  affecting sex determination, 146,
    161–162
Autosomes, 23, 40, 63, 69, 71, 73,
  76, 140, 153, 155
  and sex differentiation, 22, 25, 27,
    156, 192–193

## B

B chromosomes, 58–59, 69, 117–118,
  183
Bacteria, 134–135, 188, 195
  donor and recipient, 135
Balance, genic, 23, 25–28
Balbiani rings, 178
Banding patterns, 78, *see also* Giemsa
  staining
*Bar,* 164, 167–168
Barr bodies, 61–63, 73–74, *see also*
  Sex chromatin
Bases, purine, pyrimidine, 47, 48
  base substitution, 164
*Betta splendens,* 103
Birds, 15, 90, 105, 119–132, 163, 192,
  196, 198
Birth weights, 170, 182, 184
Bisexual, 83–84, 91
*Bithorax,* 165, 168
Bivalents, 178
Blastocysts, 90, 159
*Bobbed,* 51, 56
*Bombyx mori,* 61
*Bonellia viridis,* 39, 133–134

242